$3.95

Celestial Navigation Planning

Celestial Navigation Planning

BY LEONARD GRAY

Cornell Maritime Press

CENTREVILLE, MARYLAND

Library of Congress Cataloging in Publication Data

Gray, Leonard.
Celestial navigation planning.

Bibliography: p.
Includes index.
1. Navigation. 2. Nautical astronomy. I. Title.
VK555.G699 1984 623.89 84-45264
ISBN 0-87033-327-5

Manufactured in the United States of America

First edition

Contents

	Acknowledgments	vii
1.	The Need for Planning	3
2.	Preliminaries	11
3.	Weather	17
4.	Timekeeping at Sea	23
5.	Calculations and Errors	29
6.	Route Planning	41
7.	Preparing the Worksheets	55
8.	Daily Navigation Guides	71
9.	The Assumed Altitude Method	79
10.	Special Problems	83
	Appendices	
	A Sight-Reduction Form for H.O. 249 or H.O. 229	90
	B Sources of Publications	92
	C Sight-Reduction Methods Compared	96
	D Checklists	99
	E Formulas	101
	F Glossary	107
	G Excerpts (*Nautical Almanac* and H.O. 249, Vol. 1)	111
	Index	127

Acknowledgments

Cover and large drawings are by Bill Tinker. Small cuts are from the *Handbook of Early Advertising Art,* by Clarence P. Hornung; Dover Publications. Excerpts from the *Nautical Almanac* are courtesy of the Nautical Almanac Office, U. S. Naval Observatory. Excerpts from Publication No. 249 are courtesy of the Defense Mapping Agency.

Celestial Navigation Planning

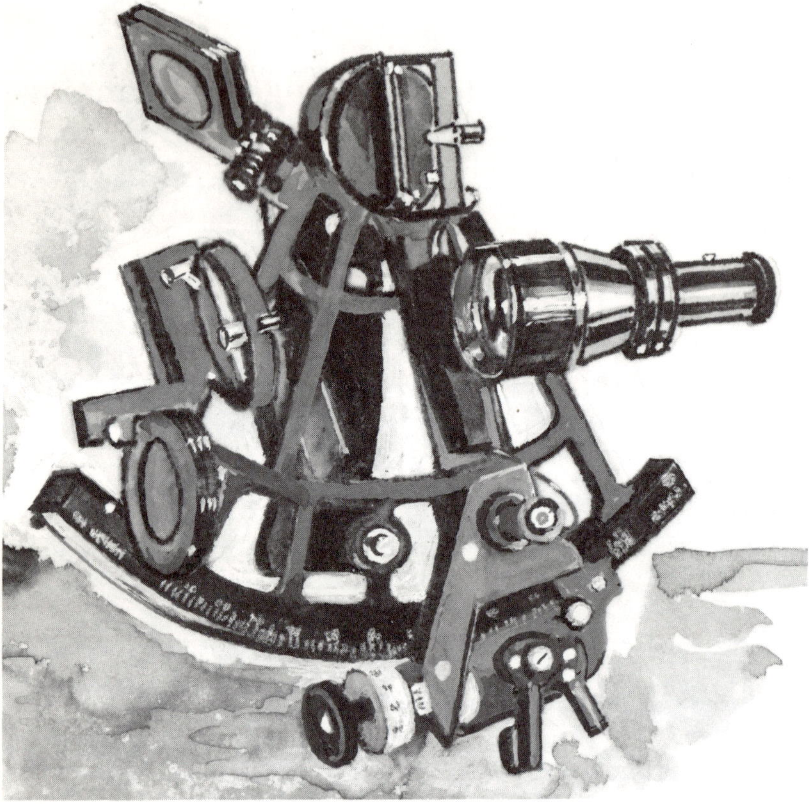

CHAPTER 1

The Need for Planning

It is 6:00 P.M. local zone time, September 2. A 39-foot yawl is in the fourth day of a cruise from Martha's Vineyard to Nassau. The Loran set stopped working 50 miles out, and only static and weak signals can be heard on the RDF. A shark ate the patent-log rotator two days ago, and no one has estimated the boat's speed and recorded it, or written down the various headings, or estimated the effect of the Gulf Stream current, so the navigator doesn't have a good DR position. The consensus of the crew is that they must be about a hundred miles off the coast, somewhere near Cape Hatteras.

The latest WWV hourly weather broadcast reported a storm with 40-knot winds and 12-foot waves centered at 29°N, 80°W, moving northeast at 15 knots, and the navigator and captain need to find the best course to avoid it. They are both a little out of practice in celestial navigation, but they do have a sextant on board, and now it's a matter of necessity.

The navigator, Pat, shoots the sun, low in the west, and decides to try some star sights at twilight. Pat again brings the sextant on deck and, as the sky darkens, shoots the moon, two stars that appear to be Arcturus and Deneb (the scattered clouds make it hard to tell), and a third fairly bright star.

When the sights are worked up, the LOPs for the moon, Arcturus, and Deneb form an impossibly large triangle, and none of the coordinates in the *Nautical Almanac* seem to match those of the unknown star. Pat struggles with the calculations for over an hour, then (feeling the pressure of the situation and being on the verge of seasickness from confinement below) makes a desperate guess, based partly on the earlier sun LOP and a last-minute estimate of speed made good, and gives the deck watch a change of course to 270°.

Early the next afternoon, the boat is under storm jib and mizzen, beating into 30-knot winds with 15-foot waves, and the crew are very uncomfortable. The course change took them into the path of the storm instead of away from it.

There was some bad luck here, but most of the trouble could have been avoided. The navigator's primary error was in failing to do some planning that would have helped when things went wrong.

Consider another navigator, Chris, in similar circumstances with the same equipment failures. Chris got a good sun-and-moon fix in the afternoon (Pat didn't even know the moon was in position for an afternoon sight), and is on deck at 2320 GMT for a round of evening star sights, plus sights of the moon, Jupiter, and Saturn, using a list of predicted altitudes and azimuths. These make the sightings fairly easy, in spite of the clouds, and Chris gets a good fix from sights of the moon, Saturn, and three stars. It confirms the DR based on the afternoon fix, and Chris is able to give the deck watch a change in course, to 110°, that avoids the worst part of the storm. What made this possible was that Chris preplanned the navigation for the cruise two weeks before leaving.

Purpose and Assumptions

This book was designed to help navigators with the complex requirements for planning an offshore cruise. The examples deal with a sailing cruise, but the system can be used just as well for powerboats. It is assumed that the reader already knows how to take celestial sights with a sextant; how to work the sights using *Sight Reduction Tables for Air*

Navigation, Pub. No. 249; *Sight Reduction Tables for Marine Navigation,* Pub. No. 229; the Davies *Sight Reduction Tables;* or another method; and that the reader is able to use the Star Finder and Identifier, H.O. 2102.

The aim is to save time at sea and improve navigation results. If some of the planning steps seem complicated and time-consuming, consider the following:

Every minute spent in preparations ashore means less time is needed for navigation at sea.

Since the work has to be done anyway, it is much better to do it in advance, ashore, where there is ample working space and time for checking and correcting errors.

It is better to be prepared (with precomputations and notes) for too many navigational events than for too few.

Some of the work is partly repetitive and, once started, goes fairly fast.

The book also includes a suggestion that the old method of sighting a rising or setting body exactly on the horizon, without a sextant, can be used in a new way to get LOPs at night, when the horizon is not visible.

Any abbreviations that are unfamiliar can be checked in the Glossary, Appendix F. Appendix B gives sources for publications cited.

Electronic Equipment

Since Loran and NAVSAT are available at reasonable prices, and Omega coverage is reasonably complete, what is the need for celestial navigation or extensive planning?

To begin with, the Loran and NAVSAT equipment could fail at inconvenient times, or the boat's wiring system could corrode and fail, or the batteries could go dead. It would be very rash of the navigator to assume that none of these things will happen. Also, Loran is not an offshore system; NAVSAT is limited to periodic fixes—often hours apart, depending on location; and Omega is fairly expensive.

Celestial navigation is still essential as a backup system for all offshore cruisers and as the primary method for those without Omega, Loran, or NAVSAT.

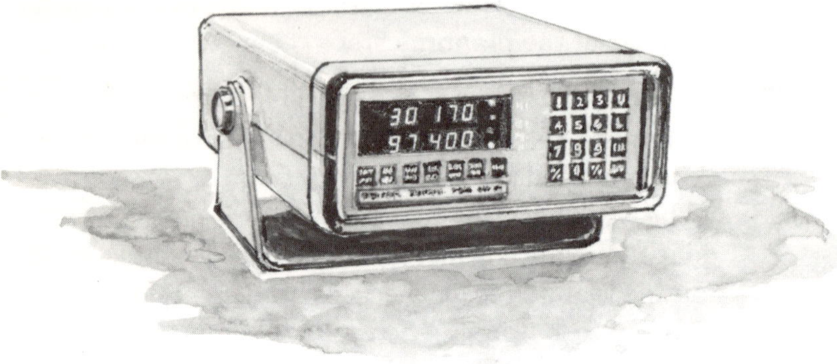

Anticipating Electronic Problems

Electronic equipment for boats has improved considerably, but is still not perfectly dependable. Sailors who have been on cruises during which only one or two—or none—of the electronic devices continued to operate might put it even more strongly. Preparations should include the following:

Take plenty of spares—dry cells, a power pack for the programmable calculator, and a second inexpensive calculator. If possible, provide two receivers capable of picking up WWV and at least some of the regular weather broadcasts.

Cover the equipment with plastic or, even better, seal each piece in a plastic bag with silica gel (available from chemical suppliers).

Provide alternatives—a slide rule for the electronic calculator; a spring-wound watch for the electronic digital watch.

Use knowledge as a substitute for equipment. If weather broadcasts can't be received, be prepared to interpret weather signs and make rough forecasts. If the patent log or electronic speed indicator has failed, throw a marker into the water at the bow, time its passage to the stern, and use the method given in Chapter 5 to find the boat's speed.

If the Loran or NAVSAT equipment fails, the navigator will need to shift to the celestial fixes already worked up as a check on the electronic fixes, supplemented by a good DR position.

Navigation Planning

Since electronics at sea are not 100 percent reliable, and celestial navigation as it is usually practiced is often too rushed, too liable to error, and too limited in scope, something more is needed. The recommended solution is (a) a system for error reduction, (b) a route plan to avoid bad weather and exploit favorable winds, and (c) compilation of the following celestial information for each day of the cruise:

Twilight times and usable bodies, with expected altitude and azimuth for each.

Times when the moon is available during the day.

Noonsight time and predicted altitude.

The time and azimuth when each available planet and the moon will be on the horizon (rising or setting) during the night.

Deviations from the Plan

What if the planning is done and then the cruise is delayed? What if the boat progresses faster or slower than the plan has contemplated?

If the cruise is delayed by a week or longer, the celestial-navigation plan should be revised, if possible; but it should not be scrapped if time is not available for recomputations. The plan can still be used, by noting deviations from the projections each day and applying the differences to the next day's projections. The same method can be used if, near the end of the cruise, the boat's progress has deviated substantially from the plan.

The solution to the second problem is to make two celestial-navigation plans—one for a speed slightly faster and the other for a speed slightly slower than anticipated. (The route planning is normally not affected by slight changes in schedule.) Much of the work for the slow plan can be used for the fast plan, so it does not take twice as long to make two plans as to make one.

The Navigator's Job

It may help to write out the navigator's daily schedule before the watch routine is planned, to facilitate the decision on which watches, if any, should be assigned to the navigator. It is strongly recommended that, if there is a full crew, the navigator not be assigned any regular watches. The navigator's schedule might look something like this (times are local boat time):

NAVIGATOR'S SCHEDULE

0500–0600	Prepare for and take twilight sights.
0600–0700	Work and plot twilight sights; establish a new compass course if required.

0800–0900	Record and evaluate weather broadcasts; get fixes by Loran, Omega, or RDF; set a new course if storms are forecast along the present course.
0900–0930	Check for compass error.
0930–1030	Take, work, and plot sun and moon sights.
1100–1130	Record and evaluate weather broadcasts.
1130–1230	Take, work, and plot the noonsight.
1400–1500	Take, work, and plot sun and moon sights.
1530–1630	Check the accuracy of the speed indicator; bring the DR plot up to date.
1700–1800	Prepare for and take twilight sights.
1800–1900	Work and plot twilight sights; set course.
2000–2100	Record and evaluate weather broadcasts; get the fixes by Loran, Omega, or RDF; set a new course.
0100–0300	Take, work, and plot star and planet sights on the horizon (see Chapter 10).

These times are only approximations, and some of the activities (such as using Loran, Omega, and the RDF) will depend on where the boat is and what equipment it has. But the list shows the other crew members that the navigator will not be loafing while everyone else is working, and must get up in the middle of the night, like the others. Sometimes—notably at evening twilight during fair weather—everyone but the navigator will be relaxing in the cockpit with a drink. When time is available, the navigator will certainly take a turn at the helm or in the galley, but should not be required to do it on schedule. The navigator's time imperatives are set by the movements of the celestial bodies and by weather-broadcast schedules—not by the watch list.

It could be said that the purpose of this book is to offer some suggestions that will help fairly competent navigators to become expert offshore navigators. (This, of course, requires experience as well as study and practice.) Aside from the important matters of safety and economy, there is a great satisfaction in accurately predicting a landfall for one's crewmates after having been out of sight of land for many days, just from squinting at the sun through a sextant, leafing through some books, scribbling some figures on paper, and drawing lines on the chart. It is especially pleasant to be able to do this when relying on sights of a couple of bodies (or maybe only one) shot hastily through the first breaks in the clouds in a week or so.

Becoming a navigation expert is recommended to anyone who is customarily stuck with the galley work and considers it onerous. But the heavy responsibility for always knowing the boat's position and for avoiding navigation dangers should not be overlooked.

CHAPTER 2

Preliminaries

Recommended Sight-Reduction Methods

Of the various sight-reduction methods available, the recommended ones are *Sight Reduction Tables for Air Navigation,* Pub. No. 249, Vol. I, Vol. II (for latitudes 0°–40°), and Vol. III (for latitudes 40°–89°). (Although Pub. No. 249 is the official name for these books, almost everyone still calls them H.O. 249, and calls Pub. No. 229, H.O. 229.)

These are two completely different methods, but they are complementary. H.O. 249 Vol. I gives computed altitudes,

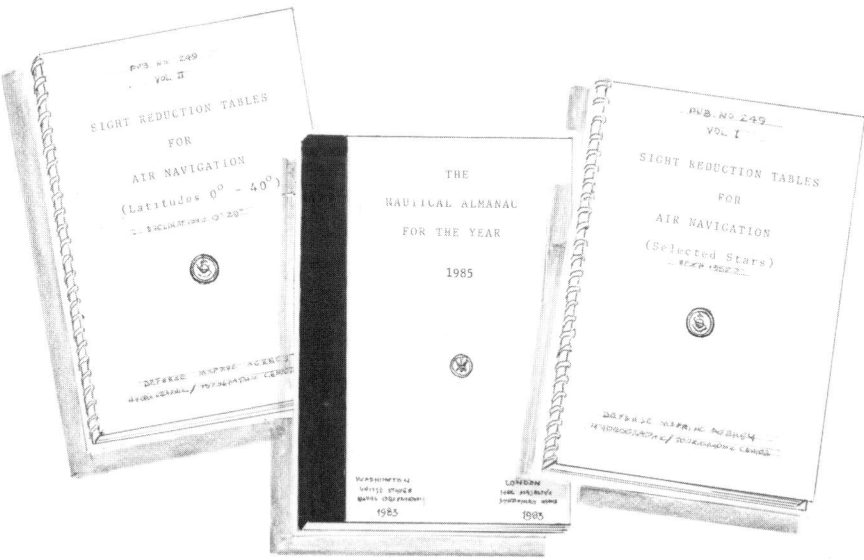

Hc, and azimuths, Zn (not just azimuth angles, Z), for seven stars found to be the best ones to sight for a given latitude and time. Entering arguments are latitude and local hour angle of Aries (LHA ♈). It is a very easy method to use for planning and identification and for sight reduction.

H.O. 249 Vols. II and III are intended for the sun, moon, and planets, and can also be used for stars with declinations less than 30°, N or S (although normally Vol. I is used for stars). They give Hc and Z (which must then be converted to Zn), with entering arguments latitude (L), declination (d), and local hour angle (LHA) of the body.

Both of these H.O. 249 methods give Hc to the nearest minute, rather than to the nearest tenth of a minute, as with H.O. 229 and some other methods. This is close enough, considering the difficulty in getting accurate sextant sights from a small boat unless the ocean is nearly calm and the sky clear. The figures for Greenwich hour angle (GHA) and declination (d) taken from the *Nautical Almanac* (or *Air Almanac* or *Reed's Almanac,* if preferred) should be rounded to the nearest minute when H.O. 249 is used. (See Chapter 5 for comments on rounding.) The gain in ease of computation—and thus the reduced probability of making a mathematical blunder—far offsets the slight loss in accuracy. There is a considerable gain in speed, too. Sextant readings (hs) should be recorded to the nearest tenth of a minute, then should be rounded to the nearest minute, for sight reduction when using H.O. 249.

A backup sight-reduction method should be available, and understood, in addition to the primary methods of H.O. 249, Vol. I, and Vols. II and III. This could be one of the compact methods—the Bayless modified H.O. 211 *Compact Sight Reduction Table* (first choice) or *Navigation Tables,* H.O. 208—or it could be H.O. 229, or a pocket calculator. Trigonometric tables are not recommended as backup (except for lifeboats)—even for math enthusiasts—unless another, faster, backup method is also on hand. (See Appendix C for comments on sight-reduction methods.)

Star and Planet Identification

With the proper planning and using H.O. 249 Vol. I for stars and H.O. 2102 for planets, identification during sighting is simple. (See Chapter 8.) It requires only that the navigator set the sextant to the precomputed altitude, point it in the precomputed direction, then look for the brightest body a few degrees up, down, left, and right—usually within, or nearly within, the field of view of the sextant telescope.

Even so, navigators should learn how to find the important navigational stars and then renew their knowledge by actually finding them—and all the visible planets—in the morning sky and evening sky, a week or so before the cruise.

Sight-Reduction Practice

A navigator who has not worked any sights for a few months should find some problems in a textbook or from another source, such as the log of a previous cruise, and work them to get old habits back into shape. A form, such as the sample in Appendix A, is recommended, so that everything will be properly labelled and nothing will be overlooked. The backup method and special techniques, such as high-altitude sun sights (see Chapter 10), should also be reviewed.

Rating Timepieces

All of the timepieces that are to be used for navigation should be rated before the cruise. The error for each one,

found from the WWV time signal or other precise source, should be recorded, along with the name and serial number, at the beginning and end of the checking period; then the rate should be computed. For digital watches, a week or so will suffice, but spring-wound watches should be checked for a month, because of their greater variability.

The data should then be summarized and recorded in the navigator's notebook:

Navigation Timepieces

		Error		
Name	*Serial No.*	*Amount*	*As of*	*Daily Rate*
Arnett	773270	Fast 2 sec.	3/17/84	+0.5 sec.
Benson	1235	Slow 9 sec.	3/17/84	−1.3 sec.
Cromwell	A16-290	Slow 4 sec.	3/17/84	−0.4 sec.

Checking the Sextants

The mirrors and shade glasses of the sextants should be carefully cleaned with soft tissue and alcohol, and adjusted if necessary. The navigator should at least check the index error and side error, sighting on a star, and adjust the instrument if there is much of the former or any of the latter. The adjustments are not difficult; explanations can be found in almost any good book on the sextant.

Cruise Preparations

Before the supplies are packed in the boat, the navigator should plan to spend a day or more checking and preparing the navigation equipment. (During the packing, the navigator should see that no ferrous metal is stowed within six feet of the compass.) Preparations should include:

Height of Eye. Measure and record the height of eye for the navigator and assistants, on deck and in the cockpit, and record these in the notebook.

One of the problems in getting sextant sights from a small boat, when the waves are large enough to lift the boat appreciably and then drop it, is related to the height of eye. In a rough sea, it is best to take each sight on the crest of a wave; this can be facilitated by having a helper call out "ready," then "now," when the boat is at the highest point. Taking the sight can be difficult, especially with dim stars. If the waves are severe, it may be necessary to start the sight on one crest, then perfect the alignment on the next crest or two.

There is also the matter of how much correction to add to the height-of-eye figure for the lifting effect of the waves. Some navigators ignore it, some add the estimated wave height, and some add half the estimated height. Adding half the crest-to-trough height is recommended as the most accurate method.

A suggestion for those who do not like to squint the left eye when looking through the sextant telescope with the right eye (or vice versa) is to try an eye patch. It can be bought in a drug store and kept in the sextant box. It is easy to use (it's held on by an elastic band) and helps relieve eye strain.

Electronic Equipment. After checking the operation of the transmitters and receivers, RDF, Loran, NAVSAT, Omega, and weather printer (if you have these), cover them with clear plastic or, preferably, seal them in a plastic bag—at least those that are used infrequently—with some silica gel inside it. This is especially important for complex items like the weather printer. Also, check the operation of the depth sounder, log or speed indicator, and steering vane or automatic helm.

Compass and RDF. If the job hasn't been done recently, swing the compass and simultaneously check the RDF with visual bearings on the transmitting antenna, so that the deviation cards can be corrected. Don't forget, when using the RDF, to move it outside the triangles formed by the shrouds and by the forestay, mast, and backstay, and to break other continuous loops (for example, by unsnapping the pelican hooks on the lifelines). Otherwise, the bearings will be affected.

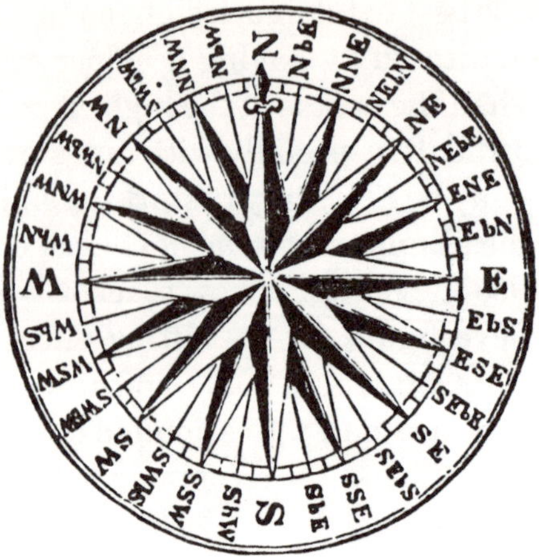

One way to check the compass is to sight past the mast and in line with the bow, using a hand-bearing compass. First, find a spot (at least six feet from the boat's compass) that is free from deviation. This can be done by taking a bearing on a distant fixed object, changing the heading 10°, and taking the bearing again, repeating for each 10° of the 360°. If the bearings are different on different headings, try other sighting locations until one free from deviation has been found. Then use the hand-bearing compass (now sighting along the mast and bow) to check the steering compass on all headings at the same time the RDF is being checked. This method has the advantage of also identifying the best location from which to take bearings with the hand-bearing compass.

Lifeboat Equipment. The backup sextant should be stowed with the emergency navigation equipment packed in the lifeboat, well protected from moisture. (If the crew is forced to abandon ship, the navigator should grab the regular sextant and any other navigation equipment at hand.) The navigator should also see that other essential navigation items are included with the lifeboat equipment. (See Appendix D for a suggested list.)

CHAPTER 3

Weather

Weather planning before and during the cruise includes:

Scheduling the cruise for the combination of route and departure time that can be expected to result in the best weather.

Compiling a list of weather broadcasts to be used during

the cruise, or annotating a copy of *Selected Worldwide Marine Weather Broadcasts.*

Waiting for hurricanes and storms to move on before starting the cruise.

Checking weather forecasts at sea and changing course to avoid storms. (See Appendix E for a storm-avoidance formula.)

Altering course around highs and lows to get favorable winds.

Information on shipping lanes and expected winds, currents, waves, and storms on various routes at different times of the year is available from pilot charts and other sources (see Chapter 6). The first step in cruise planning is to consult one or more of these references for guidance in deciding on the best time of year and the best route. For example, the rhumb line from Point A to Point B may lead through prevailing bad weather, unfavorable winds, or adverse currents at certain times of the year, and a longer route might save sailing time and increase safety.

After planning the time and the route, the cruising sailor should consult *Selected Worldwide Marine Weather Broadcasts* to determine the frequencies, call signs, locations, transmission modes, and types of information broadcast by stations that can be expected to be heard for each segment of the trip.

Morse-code broadcasts should be included if anyone in the crew knows the code. Morse code often can be heard when voice broadcasts could not be picked up at a given location. Anyone not familiar with single-sideband (SSB) voice transmissions should practice the tricky process of tuning in SSB stations before leaving on the cruise, because most weather broadcasts are in SSB. A digital frequency indicator on the receiver makes the tuning much easier.

WWV, Fort Collins, Colorado, and WWVH, Kauai, Hawaii, broadcast very useful gale warnings and other

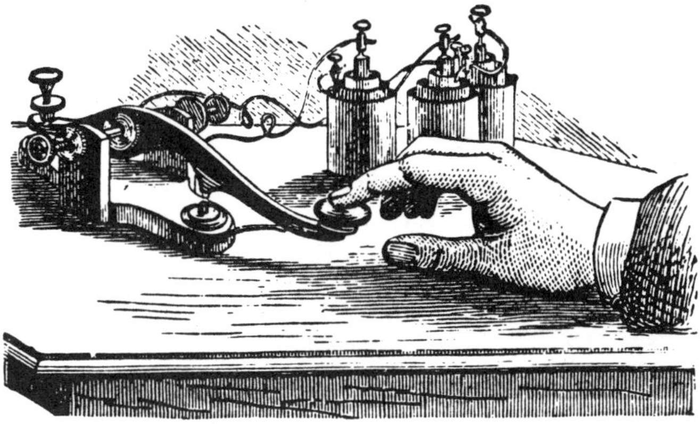

weather information on their time-signal frequencies of 2.5, 5, 10, 15, and (for WWV only) 20 MHz, just after the time signal, as follows:

Station	*Broadcast Times: Minutes after Each Hour*	*Area Covered*	*Based on Reports Compiled at*
WWV	8 and 9	Western North Atlantic, Gulf of Mexico, and Caribbean	0500, 1100, 1700, 2300 GMT
WWV	10 and 11	North Pacific east of 140° W	
WWVH	48, 49, 50, and 51	North Pacific; South Pacific to 25° S, 110° W to 160° E	0000, 0600, 1200, 1800 GMT

These broadcasts are in double sideband, from powerful transmitters, so they are easy to tune in. And, because of the constant time tick, they can be tuned in in advance. There is no need to fumble around, trying to find the station (as with SSB), while vital information is being lost. Also, if one or more of the WWV or WWVH frequencies cannot be heard at a given time and location, usually at least one of the others can be received. The announcer has to talk fast to get all the information across in the allotted time, so it pays to practice writing down the information from these broadcasts before going to sea.

While the boat is at the dock awaiting departure, the navigator should listen to weather forecasts, and should not hesitate to recommend a delay if there are any storms on the way—especially if there is a hurricane threat.

Once the cruise is under way, the navigator should immediately begin to listen to the weather broadcasts listed in the plan.

Recording Weather Broadcasts

The navigator should not try to memorize weather broadcasts. Yet the information, either in Morse code or in voice, is usually presented so fast that it is hard to record in longhand. A prepared format helps considerably, because it shows what is coming next. The following form is suggested for the storm warnings broadcast by WWV.

WWV Storm Warnings

Date	Time	Compiled At	Weather Feature	Center		GMT	Moving		Wind	Waves	Danger Radius		Remarks
				L	λ		Dir.	Speed	(Knots)	(Feet)	Miles	Quadr.	

Fig. 1. WWV weather-report form.

Note that this format is subject to change. The navigator should get the current one by listening to WWV just before the cruise. Even with a form, it may be hard to record all the data, especially if there is interference or the signal is weak. A small battery-powered recorder helps considerably. It should be connected by a plug-in wire (available from electronics stores) to the receiver, and the person recording should listen in, preferably with earphones, and write down as much as possible. Then, if the recorder has been inoperative, at least some of the message will have been preserved.

The recorder has another advantage. A person who can copy Morse code, but not very well, can record a message and then play it back repeatedly, copying a small portion at a time. A two-speed recorder makes this even easier.

One particularly useful weather broadcast is the surface-analysis message, which consists of a long series of five-digit groups that provide all the information needed to construct a weather map of a part of the North Atlantic—either southern (5°N to 15°N, 45°W to 60°W), eastern (east of 35°W, plus the Mediterranean), or western (west of 35°W, plus the Caribbean and Gulf of Mexico), depending on the station and time of broadcast. *Selected Worldwide Marine Weather Broadcasts* contains the necessary information on broadcast times and frequencies and an explanation of the numerical code.

These broadcasts can be very helpful to the crew of a boat without a weather printer, but there are two drawbacks: the broadcasts are in Morse code; and it takes several hours to copy down the numbers, translate them into weather terms, and draw a weather map from the information. However, since the broadcast is entirely in numbers, copying the Morse code is not as difficult as copying a message in plain language. It would even be possible for someone who doesn't know Morse to learn *only* the numbers (no letters), just for this purpose. Also, the navigator can tune in the station, start a tape recorder, go about other business (but check for frequency drift now and then), and replay the tape when other duties are not urgent.

Predicting Thunderstorms

Local-air-mass thunderstorms that are not part of a storm system will probably not be forecast in weather reports received at sea. Fortunately, there is a very simple and reliable method for detecting nearby thunderstorms. It is considerably more dependable than such old sayings as:

> Red sky at night, sailors' delight;
> Red sky at morning, sailors take warning.

Some of those old rules are not entirely worthless, but they are not entirely reliable, either. The following method is very useful. It requires only an ordinary AM radio—a small portable one will do—tuned to any frequency (the lower the better) in the broadcast band, but not to a station. The volume should be turned up. If there is a close thunderstorm, periodic bursts of static—a kind of electronic crashing noise—will be heard, and will be stronger and more frequent the closer the storm.

The only certainty about these indications is that there is a thunderstorm in the locality. It might pass to the side, or it could be one that has already passed over the observer. Some people have reported taking bearings on repeated bursts of static noise with an ordinary RDF, and it may be worth trying, but it is very difficult, and resolving the reciprocal ambiguity is nearly impossible. The low-frequency band on the RDF, however, is somewhat better than the broadcast band for receiving static noise. (This is another way of saying that static is more of a problem to reception at the lower frequencies.)

Any time there are cumulus clouds building up—usually in the afternoon in hot, humid weather—it pays to listen for static noise every fifteen or twenty minutes.

An aneroid barometer should be taken on the cruise, and its reading should be recorded every hour. It should be calibrated for sea level before the cruise, so that its indications can be compared to those in weather reports. This is usually done by simply turning a small screw on the back of the instrument until the needle points to the current sea-level reading obtained from a weather station.

CHAPTER 4

Timekeeping at Sea

Many lives and ships and cargoes were lost in the early days of navigation because it was not possible to determine longitude at sea. (Latitude could be found, mainly by noonsights of the sun, using the crude precursors of the sextant.) By the eighteenth century, accurate time offshore—and therefore the means of determining longitude—was so urgently needed and so far from being a reality that a Board of Longitude was established in Great Britain, in 1714, to deal with the problem. The board offered £20,000—a fortune at the time—for, in effect, the invention of an

accurate chronometer.

A self-educated former carpenter and watchmaker named John Harrison met the requirements for the prize after years of effort and the use of considerable ingenuity. (Unfortunately, he was not awarded the entire prize for his first efforts; he was an old man when he got the last of his £20,000.) The chronometer the Board of Longitude had constructed from Harrison's plans cost the considerable sum of £450.

Now we have the quartz digital watch, which is extremely accurate, light, and portable, and ridiculously cheap. (As this is written, digital watches—with few special features and not beautiful, but just as accurate as fancier models—are being sold in discount stores for $3!) The small errors of these watches can be checked daily against the time signals broadcast by WWV and other stations.

But, since nothing is foolproof (watches can be lost or broken, batteries can go dead), it pays to take some precautions. A good recommendation for navigators is to take on the cruise one spring-wound watch (for insurance) and several quartz digital watches. One of the digital watches should be set to GMT and left that way. The timepiece that is set to local boat time and changed as the time zones are crossed should not be one of the navigation watches. It could even be an alarm clock.

The reason for having three or more navigation watches is that if time signals become unavailable and one of the watches loses its accuracy, the median of the watch readings (the one in the middle) can be relied on, because the inaccurate one will have little or no effect on the median. The navigator should arrange to have at least one, and preferably two, receivers on board capable of receiving time signals from WWV/WWVH and from stations in Canada, Mexico, England, Germany, and other countries. *Radio Navigational Aids* lists these.

Local Boat Time

Some navigators work with local zone time (LZT) and convert it to GMT for celestial observations, then back to LZT for labelling plots. This is a nuisance. It is much simpler to keep at least one of the navigation watches set to GMT, in the 24-hour mode if possible (many digital watches allow this option), and tag the watch "GMT." All times are recorded as GMT for navigational purposes, and plots are so labelled.

It is better to use local time for regulating meals, watch standing, and other daily activities. The navigator should establish this time by fiat and set a clock for the crew. The best procedure is to change the clock, in one-hour increments, on entering a new time zone, but always at the same time of day (say, 2:00 P.M.), so there will be no complaints about extending or reducing someone's watch. The policy should be announced in advance and enforced by the navigator.

Length of the Day

For a cruise with a substantial easterly or westerly component, especially in high latitudes, the navigation planner should be aware of the strange fact that there are not exactly 24 hours in the day. Although the timepiece keeping GMT will continue to show 24 hours a day, the clock showing boat time will now and then record 23 hours (if the boat is travelling east) or 25 hours (if travelling west). It will do so because the navigator is setting it ahead or back as time zones are crossed.

Sailing east or west not only adds or subtracts an hour from a watch now and then; it has a minor but measurable effect on planning, too—not in occasional one-hour leaps, but gradually, a few minutes a day. For example, if a boat is headed east at 5 knots in 45° latitude, the time from a celestial event (such as evening twilight) one day to the same event the next day will be 23 hours 49 minutes—not 24 hours.

Expected Position Each Day

One of the first requirements for the planning worksheet (see Chapter 7) is to find the expected position each day at A.M. twilight, local apparent noon (LAN), and P.M. twilight. This is not quite a simple matter of multiplying 24 hours by the predicted average speed for the day and laying off the distance along the course line.

First, we are required to find, for example, our expected position at evening twilight on June 1. To get it, we need to know the number of hours since the previous day's P.M. twilight, so we can multiply that figure by average speed to get the distance travelled that day. But to get that number of hours, we need to know our position at P.M. twilight June 1, so we can figure the time of twilight. In other words, we need A to find B and B to find A. The easiest way to solve the problem is:

1. Find the boat's approximate expected position at P.M. twilight June 1, by multiplying 24 hours by the average speed and laying off the result, in the direction of the desired track, from the previous day's P.M. twilight position.
2. Use the approximate position thus found to get the time of P.M. twilight June 1.
3. With this time and the time of the previous day's twilight, figure the number of hours and minutes of the day's run. It should be a few minutes less than 24 hours if eastbound, or a few minutes more if westbound.
4. Multiply this number of hours and minutes by the average speed to arrive at the more accurate number of miles expected to be run since the previous day's P.M. twilight, and use it to get a more accurate position.

A similar process can be used to find each day's expected position at A.M. twilight and LAN. The average predicted speed could of course be different on different days, depending on what the pilot charts show about wind speed and direction and ocean currents. (See Chapter 7.)

Time-Saving Approximations

This complicated procedure of calculating the exact number of hours from one day's celestial event to the next day's can often be skipped, because the refinement is so slight that it is not worth the additional trouble. Unless the boat is fast, and sailing on a mainly easterly or westerly course in high latitudes, the planning can be done by just assuming that each day's run will be 24 hours long.

CHAPTER 5

Calculations and Errors

Pocket Calculators

Another miracle of modern electronics, along with the quartz digital watch, is the "slide-rule" or "engineering" pocket calculator, with trigonometric functions, square root, memory, and other features, which now sells for about fifteen dollars. There should be at least one on every cruising boat—sealed in a plastic bag with silica gel. Some of its uses are:

- Backup for sight reduction.
- Interpolation of twilight time, etc., in the *Nautical Almanac*.
- Figuring boat speed from time required to pass a floating object. (See "Estimating Speed," near the end of this chapter.)
- Computing distance of visibility of lights.
- Recomputing navigation plans at sea.

The inexpensive engineering-type calculator is excellent as a backup for sight reduction. But unless a calculator is specifically designed for celestial navigation, and has *Nautical Almanac* data wired into it, no time can be saved by using it as the primary method instead of hand calculations and H.O. 249, H.O. 229, or the Davies *Sight Reduction Tables.*

Intermediate between the engineering calculator and the specialized navigation calculator is the type that can be programmed by the user. Although this is no faster than H.O. 249 for ordinary sight reduction, it works faster and better for calculation of great-circle courses and distances, identification of unknown stars and planets, and sight reduction when the DR position is used as the assumed position (AP). This is because tabular methods require double interpolation for such problems, but calculators handle degrees and minutes as easily as whole degrees. (Some require conversion of minutes to decimal degrees, but that is not a difficult operation.)

Celestial-Navigation Program

The following program is for the Hewlett-Packard HP-33C and HP-33E calculators, and can be adapted for other RPN (reverse Polish notation) calculators. It works well, serves for all cases, and can be used to solve the following problems:

Sight reduction: finding computed altitude and azimuth for an identified body whose altitude has been observed.

Star and planet identification: finding declination and LHA for an unknown body whose altitude and azimuth have been observed.

Great-circle computations: finding the great-circle initial course and distance between two locations. (See Chapter 6.)

The program is shown in the following table. "Line" is the number of the program line, "Key Entry" indicates each key that must be pressed to enter the program into the calculator memory, and "Display" is the number that will appear in the calculator display if the proper entry has been made.

Entering Angles

The only special requirement for this program is that tenths of minutes of arc must be converted to seconds of arc for data entry, and vice versa for reading the computed data.

Celestial-Navigation Program
HP-33C & HP-33E

Line	Key Entry	Display	Line	Key Entry	Display
1	→H	15 6	25	RCL 6	24 6
2	COS	14 8	26	×	61
3	STO 6	23 6	27	+	51
4	R↓	22	28	STO 3	23 3
5	R↓	22	29	SIN^{-1}	15 7
6	→H	15 6	30	→H. MS	14 6
7	SIN	14 7	31	R/S	74
8	STO 2	23 2	32	RCL 1	24 1
9	R↓	22	33	ENTER	31
10	→H	15 6	34	RCL 3	24 3
11	SIN	14 7	35	×	61
12	STO 1	23 1	36	CHS	32
13	ENTER	31	37	RCL 2	24 2
14	RCL 2	24 2	38	+	51
15	×	61	39	RCL 4	24 4
16	RCL 1	24 1	40	ENTER	31
17	SIN^{-1}	15 7	41	RCL 3	24 3
18	COS	14 8	42	SIN^{-1}	15 7
19	STO 4	23 4	43	COS	14 8
20	RCL 2	24 2	44	×	61
21	SIN^{-1}	15 7	45	÷	71
22	COS	14 8	46	COS^{-1}	15 8
23	STO 5	23 5	47	GTO 00	1300
24	×	61			

This is not as difficult as it sounds. It can be done mentally; for example:

Required: Enter 48°32′.7.
Solution: Key in 48.3242. (The last two digits were obtained by multiplying mentally: .7 × 60 = 42.)

Required: Interpret a display of 36.1949.

Solution: Write down 36°19′, then figure the tenths mentally: 49 ÷ 60 = .8 (to the nearest tenth of a minute). The complete answer is 36°19′.8.

Use of the Program

Calculation of Hc and Zn

1. Enter latitude, L. (Enter S latitude as a minus number.)
2. Press ENTER key.
3. Enter declination, d. (Enter S declination as a minus number.)
4. Press ENTER key.
5. Enter local hour angle, LHA.
6. Press ENTER key.
7. Press R/S key.
8. Read Hc.
9. Press R/S key.
10. Read Z.
11. If LHA≥180°, Zn = Z.
    If LHA<180°, Zn = 360° − Z.

Example. For a sight of a known body taken at 38°44′N latitude, the declination is S18°10′.3, and LHA is 343°01′. Find Hc and Zn. Following the method described above, the entries are: 38.44, ENTER, 18.1018, CHS (for minus, or south, declination), ENTER, 343.01, ENTER, then R/S, and (after Hc is displayed) R/S again.

The first display, 30.5440, indicates 30°54′.7 for Hc; the second display, 161.1271, indicates 161° for Zn.

Identification

1. Enter latitude, L. (Enter S latitude as a minus number.)
2. Press ENTER key.
3. Enter altitude, Ho.

4. Press ENTER key.
5. Enter azimuth, Zn.
6. Press ENTER key.
7. Press R/S key.
8. Read declination, d. (Read a minus number as S declination.)
9. Press R/S key.
10. Read LHA. (If Zn $< 180°$, use 360° minus this figure.)
11. Compute LHA + λW (or − λE) − GHA♈ = SHA body.

Example. For a sight of an unknown body taken at 25°01′N latitude, the observed altitude is 47°36′.1 and the observed azimuth is 112°. Find d and LHA. Following the method described above, the entries are: 25.01, ENTER, 47.3606, ENTER, 112, ENTER, then R/S, and (after d is displayed) R/S again.

The first display, 4.4701, indicates N4°47′.0 for declination (marked N because there is no minus sign displayed). The second display, 38.8562, indicates 39°. According to the rule (since Zn $< 180°$), the LHA is 360° − 39° = 321°.

Navigation Errors

It seems contrary to nature to check and recheck computations, once they are completed. Most people, when a thing is finished, want to leave it and go on to the next project. The careful navigator must resist this impulse. Unless the navigator is sure, from having carefully checked the work, that there is not an error, the job is not finished. And if there is a serious error, the work is probably less useful than if it hadn't been done at all. Ordinary care will not do. Navigation requires extra care and constant checking.

Celestial navigation is intricate, but its individual steps are simple. Possibly for this reason, there are many opportunities for error. Everyone makes mistakes—even careful navigators—but the careful navigators correct theirs.

Small differences between measured values and true values, such as in the sextant reading and the time, and in plotting, are not the real threat. That comes from blunders. Those who think they will never read a sextant angle of 39°43′.2 as 34°43′.2 or record a digital watch reading of 2137:15 as 2137:51 should recall whether they have ever written something like January 7, 1983, in making out a bank check for January 7, 1984.

For the 99+ percent of us who have made such blunders, the only sensible thing is to realize we will do it again, and take steps to minimize the damage. Recommendations are as follows.

Check everything, preferably by a different method. Specifically:

Keep accurate DR records, taking account of estimated current and leeway, and check each fix against the DR position.

Compare observations with predictions. For example, if twilight sights were taken at a time greatly different from the predicted time, go over the figures for the prediction and the sight reduction and resolve the discrepancy.

For sun or sun-and-moon observations, take five or six sights of each body, then repeat the process an hour or so later.

Work sun or moon sights once, then work them again, without looking at the original figures. Work the noonsight as usual, then work it as a conventional LOP, using H.O. 249 or another method.

If a round of sights plots well, accept it as correct after checking the date, time, index correction, and dip correction. If one or more of the LOPs plots far from the others, rework the suspect sights (preferably using a different method) without looking at the original figures.

Rounding Off

Many people round off decimal figures by dropping the last digit if it is 4 or less and raising the previous digit if the last one is 5 or more (and similarly if dropping more than one digit). A better way, because it allows a better chance for the rounding errors to cancel out, is this: when rounding for a final digit of 5, round odd digits up; round even digits down. For example:

7.2501 rounds to 7.3, as with the more common method of rounding (because the 1 on the end makes .2501 closer to .3 than to .2).
7.3500 rounds to 7.4 (because 3 is odd).
7.8500 rounds to 7.8 (because 8 is even).

Checking for Errors by Differencing

There is a simple and effective method for quickly checking a set of figures in a series, such as those on the navigator's planning worksheet—for example, the column showing the LMT of A.M. twilight for each of several days. (The worksheet is covered in Chapter 7.) Let's assume the figures are:

A.M. Twilight

Day	*LMT*
1	0501
2	0501
3	0502
4	0510
5	0503
6	0503
7	0504

The differences from each day to the next are:

Day	*Difference (Minutes)*
1	
	0
2	
	+1
3	
	+8
4	
	−7
5	
	0
6	
	+1
7	

In the second table above, there is obviously something wrong between Days 3 and 4 and between Days 4 and 5. A glance at the first table shows that the problem is apparently in the LMT for Day 4, 0510, which suddenly jumped 8 minutes. This is the one that should be checked.

There is one thing to note, however. The tables on the right-hand daily pages of the *Nautical Almanac* that show sunrise, sunset, and twilight cover all of the three days on each page. Therefore, values taken from the table for the first or third day and used without interpolation for date will be slightly off the true value. The *Nautical Almanac* tells how to correct for this error, but it is so small that it should be ignored for most ordinary purposes and just kept in mind when checking figures by differencing.

Also, in using this method to check a table of figures for the moon, note that the increments may not always be the same; the difference may be increasing or decreasing. The differencing method is still useful, however, because there is a pattern that can be seen in the increments.

Averaging to Reduce Errors

All observations should be repeated, if possible—sun, moon, star, and planet sights, RDF bearings, estimates of speed, course made good, leeway, and everything else. Linear regression can be used to average some of these if a calculator with that capability is available, but it is usually just as easy to plot the points (for example, sextant altitudes vs. time) and draw in an averaging line. In fact, it is mathematically better—sometimes essential—to do this instead of using linear regression if the points do not plot as a straight line (for example, for a noonsight).

When averaging, it is often a good idea to discard the occasional observation whose value is far from the others. But obviously this should be done with caution if there are only a few values.

Possibilities for error are very high in navigation—especially celestial—but, unfortunately, it is not always possible to check one thing against another. This means that in many instances the only possible check is repeating the computations. Boring as it is, this is essential. The following techniques are suggested.

Checking Methods

Reduce all operations to a routine that you have practiced—at sea or ashore—many times. Use printed or duplicated forms (see Appendix A for a sample), with as many notes and instructions as you need. Check and duplicate observations as follows:

In taking a round of sights at twilight, use six or more bodies, if possible, because three could produce a fix like this:

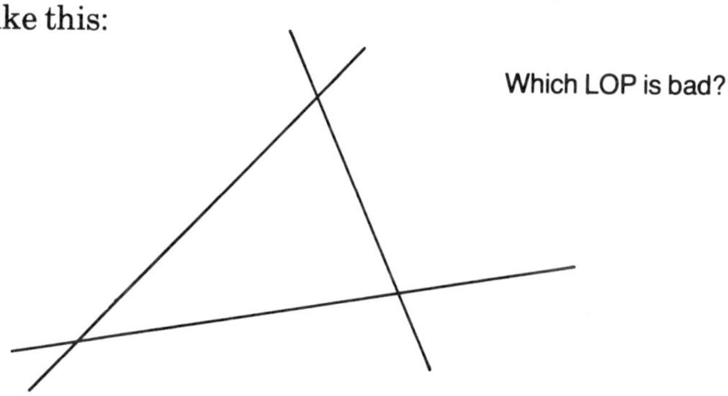

Fig. 2. Bad LOP plot.

instead of this:

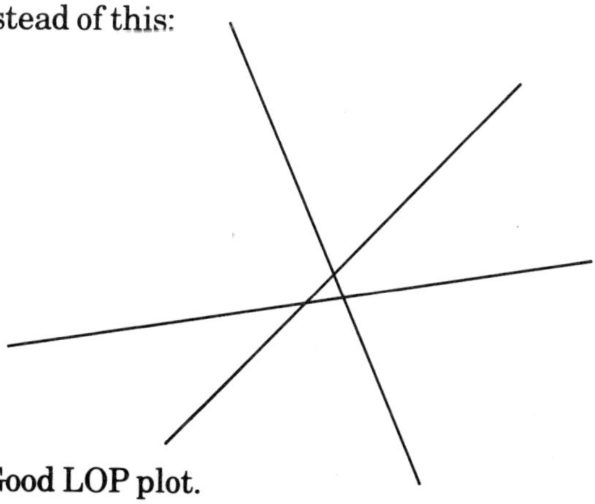

Fig. 3. Good LOP plot.

This plot, with four LOPs, is much better:

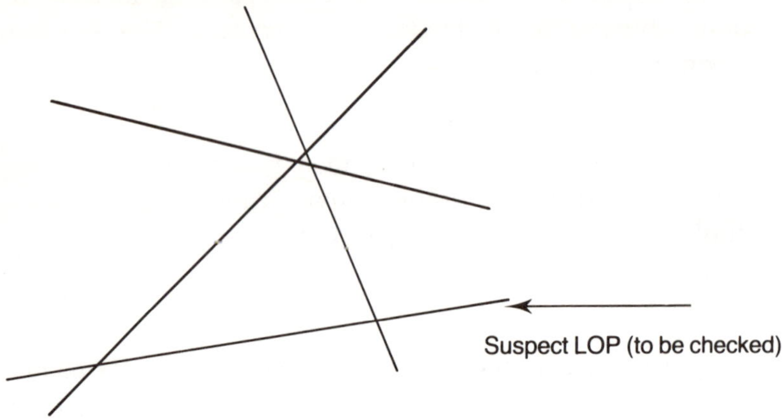

Fig. 4. Four-line LOP plot.

Now the suspect LOP is apparent and can be singled out for checking. Five or six LOPs would give even more confidence.

For sun or sun-and-moon sights, take five or six shots of each body and reduce and plot at least two each. If the results don't look right, reduce and plot the others.

If only one sight can be taken for some reason (such as cloudy weather), work it twice, by two different methods, and go through the plotting twice.

Check each celestial fix against the DR position and, if available, a fix or LOP by electronic means.

Dead Reckoning

It may not seem important to keep a good DR plot out in the middle of the ocean, but there are several reasons why this ought to be done. Every fix, celestial or otherwise, should be evaluated to see if it makes sense. Also, it is never certain that celestial sights will be available tomorrow and the day after, and a DR position may be all that is available.

An accurate position, even when the boat is far from

land, may suddenly become very important if it is necessary to maneuver to avoid a storm, assist someone in a nearby boat, or send a distress message.

Estimating Average Heading

At the end of each hour, someone should estimate the average heading steered—which may be different from the one that was desired—and record it in the log. This should be done even if a steering vane or automatic helm was steering. At night, it is often easier to get the boat on course and then pick a star for a steering guide than it is to stare at the lighted compass, even though the star must be changed periodically. Whether or not this is done, it may be better to have another person on watch (if there is more than one) estimate the average heading.

Estimating Leeway

If the boat is close-hauled, the leeway angle should be measured each hour, or on each tack, and entered in the log. One way to do this is to stand at the stern and take a bearing on the boat's wake with a hand-bearing compass.

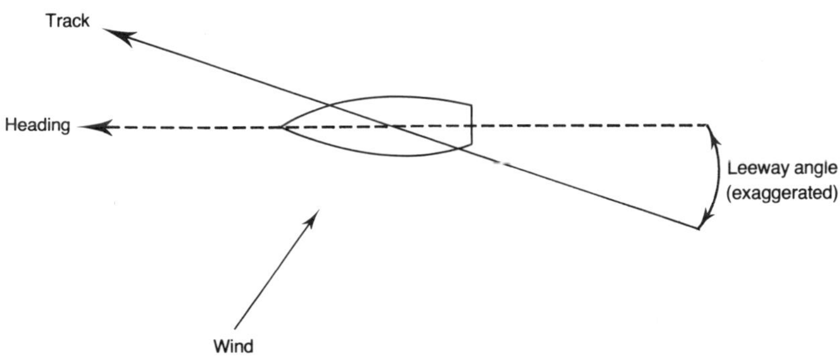

Fig. 5. Leeway diagram.

Estimating Speed

Even if there is no speed-measuring device in the boat, the speed can be measured fairly accurately by timing the passage of a floating object thrown overboard at the bow. The formula is:

$$\text{Knots} = \frac{.592 \times \text{feet}}{\text{seconds}}$$

For "feet," it is best to use the waterline length of the boat. This means that the floating object must be thrown to hit the water in front of the forward-most part of the waterline length. The timing starts when the part of the bow at the waterline reaches the object. A helper is needed in the stern to signal the person in the bow when the object is just opposite the aftermost part of the waterline length. For example:

LWL: 35 feet

Object passes in 3.6 sec.

$$\text{Speed} = \frac{.592 \times 35}{3.6} = 5.8 \text{ knots}$$

CHAPTER 6

Route Planning

Equipment

The following equipment, or its equivalent, is needed for navigation planning:

Pilot charts; also, if desired, *Ocean Passages for the World.*

H.O. 249 Vol. I (unless the Davies *Sight Reduction Tables* are to be used).

Star Finder and Identifier, H.O. 2102 (Rude's).

The *Nautical Almanac* or, if preferred, the *Air Almanac* or *Reed's Almanac.*

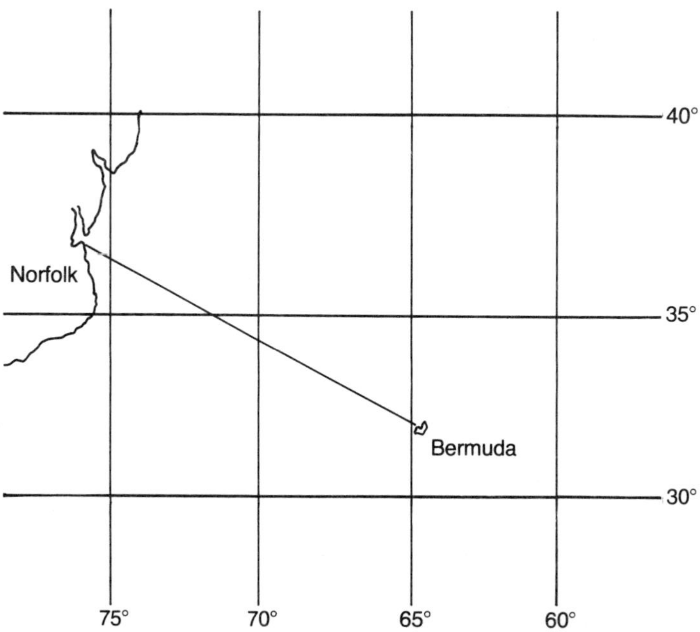

Fig. 6. Norfolk-to-Bermuda chart.

Worksheets (see Chapter 7).

Calculator. This should include trigonometric functions and storage capacity. A programmable model makes things easier.

Mercator chart or plotting chart.

Protractor or parallel ruler, dividers, pencils, eraser.

First Essentials

The first particulars the navigation planner will need are: departure point, destination, the range of acceptable departure dates, and a knowledge of the boat and its equipment.

The first determinations the navigator must make are: optimum departure time, or range of times, considering speed and safety; best route; and distance.

To illustrate the details of route planning, we will use as an example a cruise from Lands End to New York. We will choose the route and departure date after studying the *Pilot Charts of the North Atlantic Ocean.*

Pilot Charts

There is one chart for each month, giving information on winds, currents, waves, gales, hurricanes, air and sea temperatures, steamship routes, and other narrative and pictorial data. Average winds are shown, for each 5° latitude by 5° longitude block of the pilot chart, by strength and frequency (percent of total) for each of the eight points of the compass. The percent of time during which gales have been recorded is also shown for each 5° by 5° block. Currents are indicated by arrows and figures throughout the chart, and hurricane paths are depicted for the months with hurricanes. Instructions for interpretation are included on each chart.

Analysis of the pilot charts can be simplified by making up a table like the following one, which compares the direct route from Lands End to New York with a southern one, starting SSW, turning W at about 25° latitude, and continuing NW to New York.

A graph of the data for gales and waves shows the trends at a glance. July has the fewest gales—practically none—and the least severe waves for either route. It is apparent that the danger of hurricanes could be avoided by making the passage between December and April, but only at the cost of a greatly increased probability of gales.

Lands End to New York
Direct Route—Upper Figures
Southern Route—Lower Figures

Month	*Miles*	*Waves over 12 Feet*	*Hurricanes*	*Gales*	*Favorable Winds*	*Currents in Knots*
April	2900	20%	No	7%	40%	0.5 adverse
	4200	5%	No	3%	65%	0.4 favorable
May	2900	10%	Starting	3%	40%	0.5 adverse
	4200	3%	Starting	2%	65%	0.4 favorable
June	2900	5%	Yes	1%	40%	0.5 adverse
	4200	1%	Yes	0	65%	0.4 favorable
July	2900	2%	Yes	0	40%	0.5 adverse
	4200	0%	Yes	0	65%	0.4 favorable
Aug.	2900	5%	Yes	1%	40%	0.5 adverse
	4200	3%	Yes	0	65%	0.4 favorable
Sept.	2900	12%	Yes	3%	40%	0.5 adverse
	4200	5%	Yes	1%	65%	0.4 favorable
Oct.	2900	20%	Yes	6%	40%	0.5 adverse
	4200	10%	Yes	3%	65%	0.4 favorable
Nov.	2900	30%	Ending	9%	40%	0.5 adverse
	4200	15%	Ending	5%	65%	0.4 favorable
Dec.	2900	35%	No	11%	40%	0.5 adverse
	4200	15%	No	6%	65%	0.4 favorable
Jan.	2900	40%	No	13%	40%	0.5 adverse
	4200	15%	No	7%	65%	0.4 favorable
Feb.	2900	37%	No	13%	40%	0.5 adverse
	4200	12%	No	6%	65%	0.4 favorable
March	2900	27%	No	10%	40%	0.5 adverse
	4200	10%	No	4%	65%	0.4 favorable

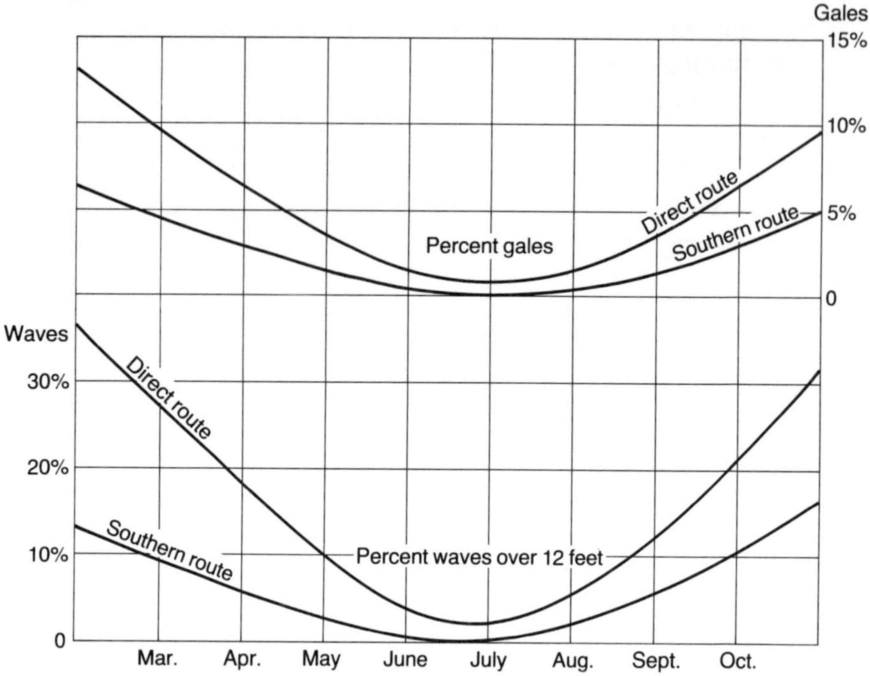

Fig. 7. Lands End to New York, direct route vs. southern route.

Best Route

A study of the wind roses shows that more favorable winds prevail on the southern route than on the direct route—something like 65 percent favorable (southern) compared to only 40 percent favorable (direct). Regardless of the time of year, the southern route has fewer gales and smaller waves than the direct route, although the differences are less in the summer than at other times.

The preference is definitely the southern route and a departure in June, July, or August. *Ocean Passages for the World* confirms this route choice in Chart 5308, "The World Sailing Ship Routes." There is, however, a danger of hurricanes during the summer for the last part of the cruise, and this must be considered along with the lower frequency of gales in summer. If this cruise is made in summer, the navigator must listen to weather reports regularly, and be prepared to make a drastic course change as the boat approaches the mainland.

Norfolk-to-Bermuda Example. Our next example demonstrates the entire planning process. Assume that we are planning a cruise from Norfolk to Bermuda, for the spring or summer of 1985, in a typical 38-foot sloop.

We can see from the pilot charts that for this cruise there is no advantage in deviating from a direct route in an attempt to find more favorable winds and currents. But we do need to find the best departure date.

The data from the pilot charts can be summarized as follows:

First Segment
35°–40°N, 70°–75°W

		April	*May*	*June*
Gales		3%	1%	0
Hurricanes		No	Starting	Yes
Winds	Favorable (NE–SW)	67% Force 4	62% Force 4	56% Force 4
Winds	Unfavorable (E–S)	30% Force 4	34% Force 4	40% Force 4
Calms		3%	4%	4%
Waves over 12 ft.		10%	5%	0

Second Segment
30°–35°N, 65°–70°W

		April	*May*	*June*
Gales		2%	0	0
Hurricanes		No	Starting	Yes
Winds	Favorable (NE–SW)	68% Force 4	54% Force 4	51% Force 4
Winds	Unfavorable (E–S)	30% Force 4	43% Force 4	46% Force 4
Calms		2%	3%	3%
Waves over 12 ft.		10%	5%	0

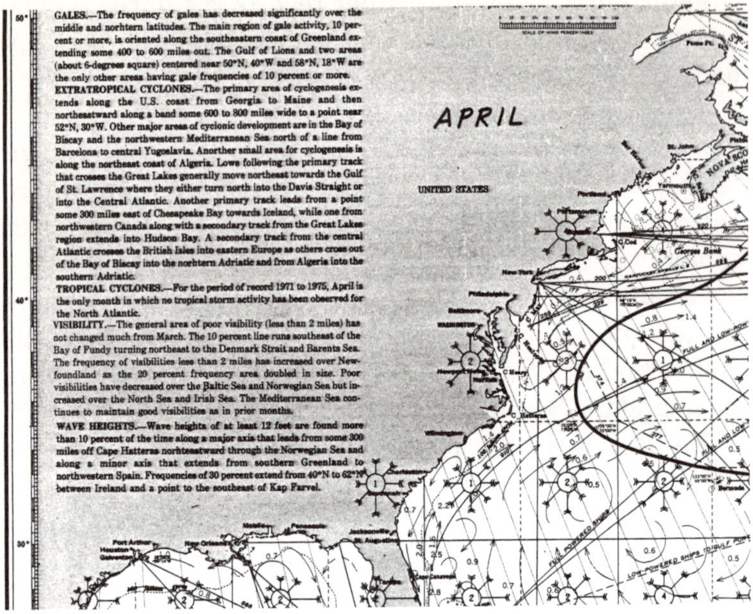

Fig. 8. Pilot chart for April (excerpt).

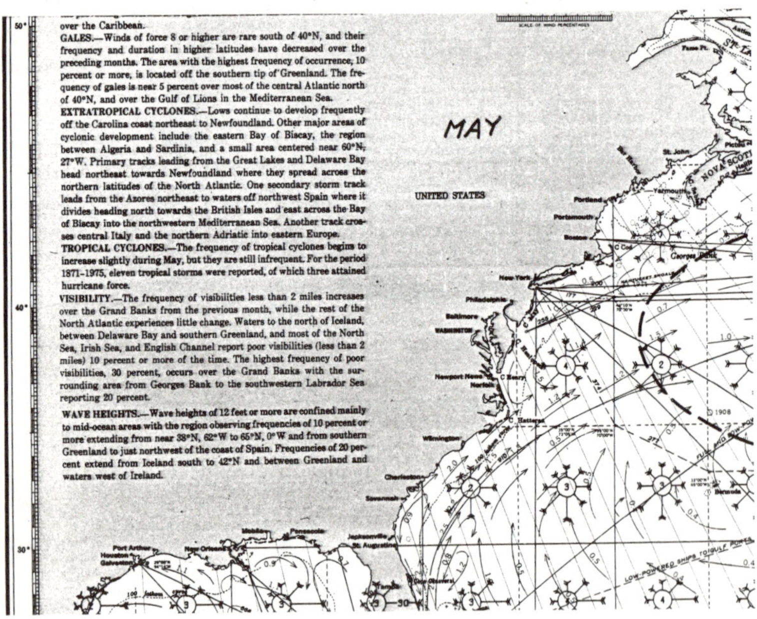

Fig. 9. Pilot chart for May (excerpt).

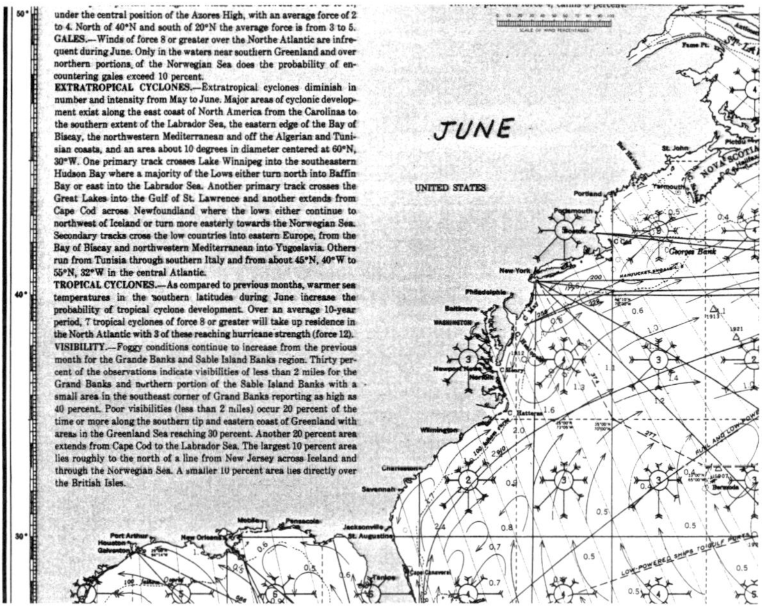

Fig. 10. Pilot chart for June (excerpt).

April vs. May Departure

Gales slightly more frequent (2–3 percent vs. 0–1 percent).
Hurricanes rare (usually starting in May).
Winds more favorable (68 percent vs. 58 percent).
Calms slightly less frequent (2–3 percent vs. 3–4 percent).
Greater probability of waves over 12 feet (10 percent vs. 5 percent).

Best Departure Date

Considering everything, the best time to leave Norfolk is around the middle of April. Although June has the advantage of having practically no gales or high waves, it has favorable winds only a little more than half the time (compared to two-thirds for April), and has slightly more calms than April has. Also, the hurricane season is under way by June. In April, the probability of gales, while greater than for June, is only 2 to 3 percent, and there is very little chance of hurricanes. The prevailing winds in April are Force 4 (11 to 16 knots) two-thirds of the time, and waves are over 12 feet 10 percent of the time. If we plan to leave in mid-April and the trip is delayed a week or two, conditions will still be favorable.

Predicted Currents

Now that we have our route and departure time, we need to estimate from the pilot charts the current speeds and directions we will experience in various stages of the cruise. Instead of doing this in minute detail, it is usually better to do some combining to reduce the amount of data to manageable proportions. A table can be made up, as follows:

In the first column, list each 2° band of longitude (since the cruise is mostly easterly).

From the pilot chart for April, list the set and drift of the current for each of these longitude bands. In some instances this will require averaging two or more current arrows.

The table will be used to get each day's course and predicted speed.

Predicted Currents
From April Pilot Chart

Longitude	*Set*	*Drift*
76°–74°	145°	0.3
74°–72°	045°	1.2
72°–70°	050°	1.1
70°–68°	050°	0.7
68°–66°	190°	0.5
66°–64°	190°	0.5

Average Speed

It is not quite correct, mathematically, to average two or more speeds by taking their sum and dividing by the number of speeds averaged. This is because the slower speeds affect the average more than the fast ones—because more time is spent at the slower speeds—and, to be precise, the formula should reflect this. (The mathematically correct formula for averaging speeds is the one for the weighted harmonic mean: $V = \Sigma d \div \Sigma [d \div v]$, where d = distance for a portion and v = the speed for that portion.) However, it is accurate enough, for computing expected speed made good (SMG) for planning purposes, to use a simple average of the projected-current vectors for each relatively small portion of the cruise—say, 2° of longitude, as we are doing here—and use that to figure the expected SMG for that portion. (See Chapter 7 for SMG computations.)

Great-Circle Sailing

For most cruises, or legs of cruises, the course should be laid out as a rhumb line. It is only when one or more parts of the cruise are in relatively high latitudes and the course is close to easterly or westerly that it pays to lay out the course on a gnomonic chart (on which a great circle plots as a straight line) and then transfer it to a Mercator chart in segments—or to do this mathematically.

Since we will not be sailing in high latitudes, there are no land obstacles in our path, and there will be no problems with icebergs (but note that icebergs surprisingly get as far south as 40°N in July), we will lay out a direct rhumb line from Cape Henry Light (36°56′N, 76°00′W) to Gibbs Hill Light (32°15′N, 64°50′W) for our exercise.

To illustrate the method, however, we can compute the distance and initial course by great-circle sailing and compare them with the distance and course by rhumb line. The great-circle problem can be solved using one of the sight-reduction methods, such as H.O. 211, or by computation, as described in Bowditch and other reference books. H.O. 249 Vol. II cannot be used for this great-circle problem because its range of tabulated declinations (for which we substitute latitude of destination for the great-circle solution) is not sufficient. H.O. 229 could be used—it covers all declinations—but a disadvantage of both H.O. 229 and H.O. 249

Vols. II and III for great-circle computations is that a complicated interpolation for "latitude" (latitude of origin) is required. Although H.O. 211 could be used without troublesome double interpolation, it is a long, complicated method. The easiest solution is by programmable calculator, using a program similar to the one given in Chapter 5, or an engineering calculator, using the sine-cosine formula (see Appendix E).

Great-Circle Solution by Calculator

	Actual Parameter	*Renamed for Computations*	*Value*
Given	Latitude of origin	"L"	36°56′N
	Latitude of destination	"d"	32°15′N
	Difference in longitude	"LHA"	11°10′
Computed		"Hc"	79°41′.6
	Distance	90° – "Hc"	10°18′.4 (618 mi.)
	Initial course angle	"Z"	114°

When the rhumb line is plotted on a Mercator chart, it is found to be 117°, 620 miles, compared to 114°, 618 miles for the great-circle course. Therefore, there is no need to use a great-circle course; the rhumb line is satisfactory.

Estimating Average Speed

In estimating the average boat speed for planning, we need to make some assumptions, based on information in the pilot charts and knowledge of the boat we will be sailing in. The first assumption is that we will be sailing on a reach or a run for a certain percentage of the time and tacking the remainder of the time. We can estimate these percentages from information on the pilot charts.

The detrimental effect of tacking on speed and distance made good toward the destination can be estimated from the expected wind velocity; the boat's speed, close-hauled, at that wind velocity; and the closest angle the boat can sail to the wind direction. We will start with some assumptions for this example, avoiding excessive optimism. Consider a boat making progress toward a destination in the direction 090°, against a wind blowing directly from 090°.

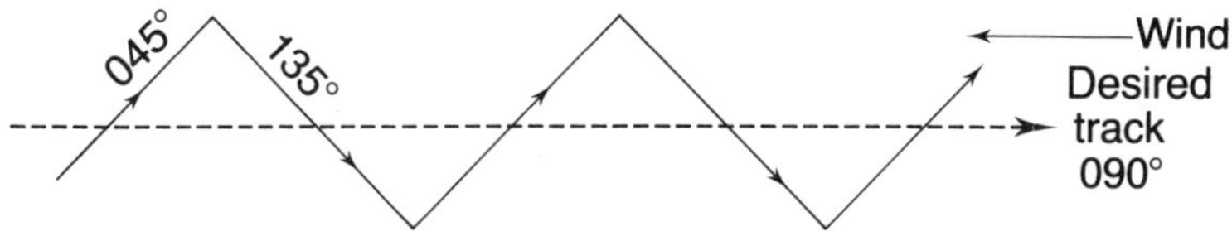

Fig. 11. Track when tacking.

Assume that the boat can sail no closer than 45° to the wind. The boat's speed, close-hauled, in a Force 4 wind, let's say, is 5.0 knots along its actual track (alternating 045° and 135° courses). The speed made good in the desired direction, 090°, is computed as follows:

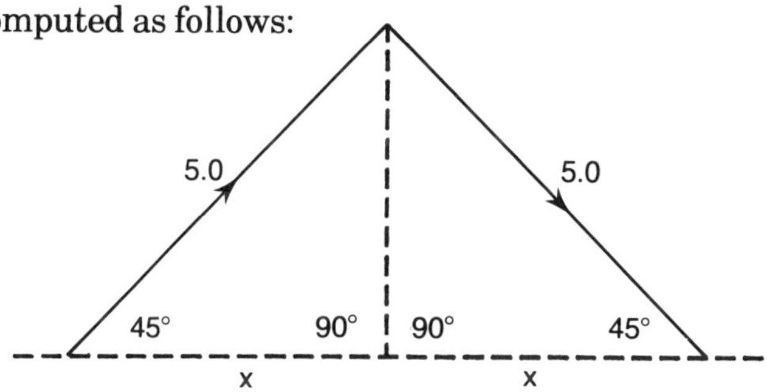

Fig. 12. SMG triangle.

$$\text{Cos } 45^\circ = \frac{x}{5.0}$$

$$x = 0.707 \times 5.0$$

$$x = 3.5 \text{ knots made good in the desired direction, } 090^\circ$$

For a boat that can sail 45° to the wind, the multiplying factor to get the SMG in the desired direction, when tacking, is 0.707 (to be multiplied by the boat's speed, as above). Factors for other angles (the cosine of each angle) are:

SMG Factors

Angle	*Cosine*
55°	0.574
50°	0.643
45°	0.707
40°	0.766
35°	0.819

The distance made good in the desired direction, 090°, is in the same proportion to the distance along the actual track (045° and 135°, alternating) as the speed made good is to the speed along the actual track. That is, for every 5.0 miles the boat travels on courses 045° and 135°, it covers 3.5 miles in the desired direction, 090°.

Now we need to figure our average speed for the conditions indicated on the pilot charts for our cruise, based on the knowledge that we can make 5.0 knots sailing free and the equivalent of 3.5 knots along the desired track when we are tacking back and forth across the desired track. (This ignores the effect of current, which is considered in Chapter 7.)

We found from the pilot charts for this cruise that we could expect to tack about 30 percent of the time and sail free about 68 percent of the time, with calms about 2 percent of the time, so we compute our average speed for the cruise as follows:

$$\text{Average speed} = (.68 \times 5.0) + (.30 \times 3.5) = 4.45 \text{ knots.}$$

For this example, then, our net average speed along our desired track should be about 4.45 knots. (Note that if we used the more precise formula, $V = \Sigma d \div \Sigma [d \div v]$, we would get $V = [.68 + .30] \div [(.68 \div 5) + (.30 \div 3.5)] = 4.42$, which is very close to the 4.45 we got using the simpler formula.)

We will make up two plans: one for a slower speed (3.5 knots) and one for a faster speed (5.0 knots). If the actual speed turns out to be close to 4.0 or 4.5 knots average, we can interpolate halfway between the two plans to get sighting times, positions of bodies, etc. If we have bad luck and average only 3.0 knots, we can still use the 3.5-knot plan with only negligible discrepancies; if we average 6.0 knots, we can use the 5.0-knot plan.

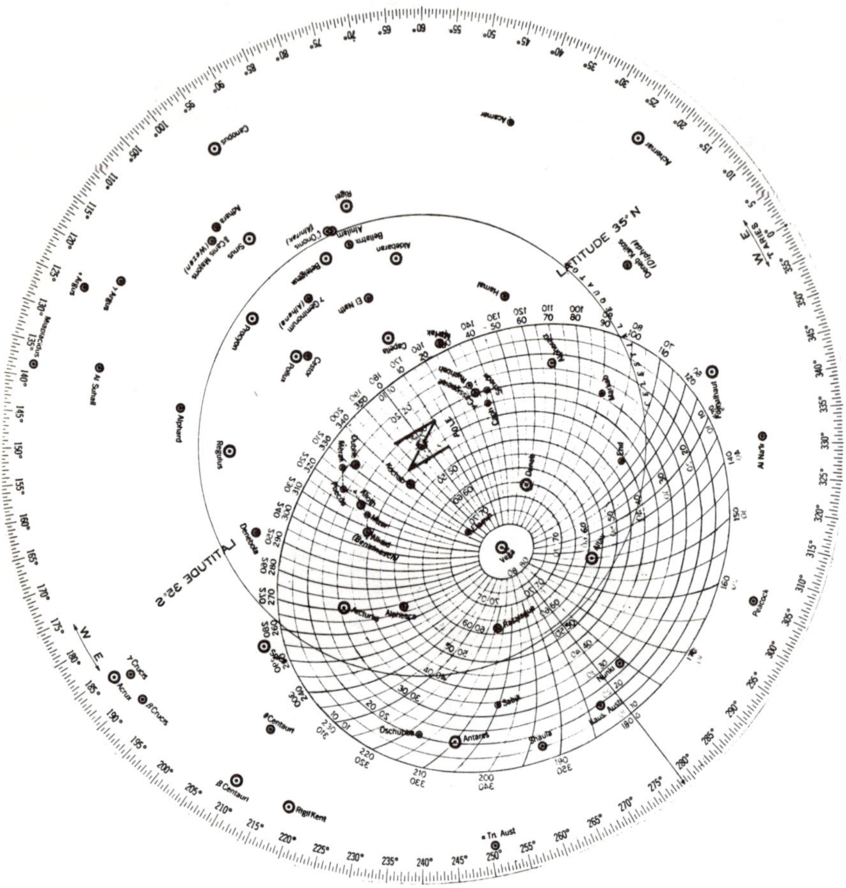

Fig. 13. Star Finder and Identifier (Rude's).

CHAPTER 7

Preparing the Worksheets

Those who work the examples in this book should expect slight differences in results, because of different degrees of precision in interpolating and rounding off. Anyone who happens to be planning a cruise from Norfolk to Bermuda should work out the plan independently, instead of using the one included here, because some of the factors—boat speed, for example—will almost certainly be different from the ones assumed for these examples.

Shortcuts

Several planning simplifications can be used to save time. A great-circle course is rarely needed; usually, as in this example, the rhumb line is accurate enough. The complication noted in Chapter 4, under "Length of the Day," and described in this chapter, is given for explanation, but is seldom needed in practice. Other useful shortcuts will become apparent as the work progresses.

Steps in Worksheet Preparation

The worksheet for the "slow plan" will be explained. The one for the "fast plan" is similar, and does not require repeating everything that was done for the slow plan.

First we will lay out the course on a Mercator or plotting chart, from Cape Henry Light (36°56′N, 76°00′W) to Gibbs Hill Light (32°15′N, 64°50′W). The desired track is 117° and the distance is 620 miles. (See Chapter 6.) Our planned departure from Cape Henry Light is at 1500 GMT, April 15, 1985.

Allowing for Current

The first step is to use a triangle of forces to find the expected speed made good (SMG) and the true course (TC) to steer for each 2° of longitude along the course, assuming a boat speed of 3.5 knots and desired track of 117°. The table that we made up showing the current for each 2° of longitude (see Chapter 6) gives 145°, 0.3 knots for the first stage, 76°–74° longitude.

For those who don't remember the triangle of forces:

1. Using a convenient scale, draw the current arrow, 0.3 units long, toward 145°.
2. From the *tail* of the current arrow, draw a long line toward 117° for the desired track.
3. Scribe an arc, radius 3.5 units, from the *head* of the current arrow, for the boat speed.
4. Also from the head of the current arrow, draw a line to the point where the arc crosses the desired-track line. Measure the angle of this course-to-steer line: 115°.
5. Measure the length of the SMG line: 3.8 units.

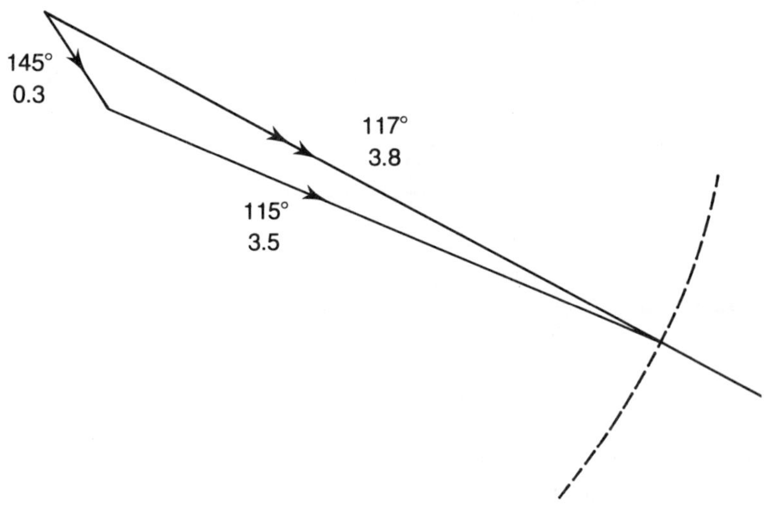

Fig. 14. Triangle of forces—course and speed.

We record the expected SMG (3.8 knots) and TC to steer (115°) for 76°–74° longitude in a new table, and then work a triangle of forces for each of the other 2° portions of longitude and complete the table.

SMG and TC to Steer for Each 2° Longitude

	Current		*From Triangle of Forces*	
Longitude	*Set*	*Drift*	*SMG*	*TC to Steer*
76°–74°	145°	0.3	3.8	115°
74°–72°	045°	1.2	3.7	136°
72°–70°	050°	1.1	3.8	134°
70°–68°	050°	0.7	3.7	128°
68°–66°	190°	0.5	3.6	109°
66°–64°	190°	0.5	3.6	109°

For the First Day

Now we are ready to make up the worksheet for the slow plan.

Our departure is at 1500 GMT, so the first celestial event will be LAN on the departure date, April 15, 1985. We will make an assumption of our longitude at the time of LAN—76°00′W—and convert this to time, using the first yellow page in the *Nautical Almanac:* 0504. The equation of time is 0, so this (1200 + 0504 + 0) provides a predicted LAN time of 1704. (See "Noonsight Projections," Chapter 10.) In 2 hours 4 minutes, at 3.8 knots, we will have gone 7.9 miles. Plotting this on the chart shows that we would be at 36°52′N, 75°52′W at LAN, and we enter these figures and the time of LAN (1704) lightly on the worksheet. To refine our estimate of the time of LAN, now that we know the expected longitude at LAN, we convert 75° 52′ to time units, getting 0503; so the more exact time for LAN is 1703. We substitute this figure for the original rough one on the worksheet.

Now we need to find the data for the next event: P.M. twilight, April 15. A quick inspection of the *Nautical Almanac* daily page shows that the LMT of civil twilight for

CAPE HENRY LIGHT TO GIBBS HILL LIGHT
SLOW PLAN
3.5 KNOTS BOAT SPEED
APRIL 15 — 22, 1985

DATE	EXPECTED POSITION AT A.M. CIVIL TWILIGHT		A.M. NAUTICAL TWILIGHT		A.M. CIVIL TWILIGHT		SUNRISE		L A N				EXPECTED POSITION AT P.M. CIVIL TWILIGHT		SUNSET		P.M. CIVIL TWILIGHT		P.M. NAUTICAL TWILIGHT		DATE
									EXPECTED POSITION		Eq. T.	GMT									
	L	λ	LMT	GMT	LMT	GMT	LMT	GMT	L	λ			L	λ	LMT	GMT	LMT	GMT	LMT	GMT	
15	—	—	—	—	—	—	—	—	36°52′	75°52′	0	1703	36°40′	75°23′	1834	2336	1900	0002 4/16	1932	0034 4/16	15
16	36°23′	74°41′	0426	0925	0458	0957	0524	1023	36°11′	74°12′	0	1657	35°59′	73°43′	1835	2330	1902	2357	1933	0028 4/17	16
17	35°42′	73°03′	0427	0919	0459	0951	0525	1017	35°30′	72°35′	0	1650	35°18′	72°06′	1834	2322	1901	2349	1932	0020 4/18	17
18	35°01′	71°24′	0427	0913	0458	0944	0524	1010	34°49′	70°56′	−1	1643	34°37′	70°27′	1834	2316	1901	2343	1932	0014 4/19	18
19	34°19′	69°45′	0427	0906	0458	0937	0524	1003	34°08′	69°20′	−1	1636	33°57′	68°52′	1834	2309	1900	2335	1931	0006 4/20	19
20	33°40′	68°12′	0427	0900	0458	0931	0524	0957	33°28′	67°44′	−1	1630	33°18′	67°19′	1834	2303	1900	2329	1931	0000 4/21	20
21	33°01′	66°41′	0428	0855	0459	0926	0524	0951	32°50′	66°14′	−1	1624	32°39′	65°47′	1833	2256	1859	2322	1929	2352	21
22	32°23′	65°09′	0429	0850	0500	0921	0525	0946	—	—	—	—	—	—	—	—	—	—	—	—	22

DEPARTURE: 1500 GMT APRIL 15

ARRIVAL: 1420 GMT APRIL 22

Fig. 15. Worksheet p. 1.

our latitude here, about 36°40′N (precision is not needed at this stage), is about 1900. Adding the time equivalent of the longitude (about 75°20′), 0501, we get a GMT of 2401—that is, 0001, April 16. We record the 1900 and 0001 lightly on the worksheet.

In the 9 hours 1 minute from departure to P.M. civil twilight, at 3.8 knots, we will have sailed 34 miles, so we plot this point on the chart. This puts us at 36°40′N, 75°23′W, which we record permanently on the worksheet. Interpolating in the *Nautical Almanac,* we find the LMT at civil twilight at 36°40′N to be 1900 (what we got in the quick estimate). Adding 0502 for the time equivalent of 75°23′ gives a GMT of 0002 (April 16) for civil twilight. We record these figures permanently on the worksheet.

In a similar manner we now get the LMT and GMT of sunset and nautical twilight, using the position of 36°40′N, 75°23′W for both. Although this is not precisely correct (we would not be at exactly the same position at sunset, civil twilight, and nautical twilight), the slight improvement in accuracy from using the exact position for each event would not be worth the extra effort. The figures are 1834 LMT and 2336 GMT for sunset, 1932 LMT and 0034 GMT (April 16) for nautical twilight.

Interpolation in the *Nautical Almanac*

There are several ways to interpolate for the exact latitude when looking up times of twilight, etc., in the *Nautical Almanac.* One option is linear regression on a calculator (see Appendix E). A second method is the use of *Nautical Almanac* Table I, p. xxxii. A third choice is illustrated by the following example:

Time of Sunset, April 15, 1985

	Latitude	*LMT*
Tabulated in *Nautical Almanac*	N40°	1837
To be found	36°40′	(?)
Tabulated in *Nautical Almanac*	N35°	1832

The difference between 35° and 36°40′ is to the difference between 35° and 40° as the difference between 1832 and (?) is to the difference between 1832 and 1837. Or, mathematically:

$$\frac{1°40'}{5°} = \frac{\text{x minutes}}{\text{5 minutes}}$$

$$\frac{100'}{300'} = \frac{x}{5}$$

$$300x = 500$$

$$x = 1.7\text{, or (rounded) 2 minutes}$$

$$1832 + 2 = 1834\text{, the LMT of sunset at } 36°40'\text{N.}$$

Another method of interpolating is to just do it roughly by eye, and this often works well enough for those who are good estimators.

Now we proceed to A.M. twilight, April 16. As before, we use a rough first guess at our position (say, 36°20′N, 74°35′W) to get an approximate time for civil twilight. This turns out to be 0458 LMT, 0956 GMT. The time from the previous P.M. civil twilight (0002 GMT) to A.M. civil twilight April 16 (about 0956 GMT) is 9 hours 54 minutes. At 3.8 knots, this would locate us 38 miles away, which plots at 36°23′N, 74°41′W. Refining the original estimate of the time of A.M. civil twilight at this more exact position, we get 0458 LMT (as in the estimate) and 0957 GMT. Nautical twilight is at 0426 LMT, 0925 GMT, and sunrise is at 0524 LMT, 1023 GMT.

After the First Day

Now the computations get easier and the worksheet proceeds much faster. For each event—A.M. twilight, LAN, and P.M. twilight—each day, we will base our estimate of position on the estimated distance run since the previous day. Also, now that the more precise method has been demonstrated, from now on we will use the simplification of assuming that this time is 24 hours, instead of using the more laborious method of finding the exact period of time, since in this instance it adds little to the accuracy.

The next event is LAN April 16. The estimated distance travelled since LAN the previous day is 24 × 3.8 = 91 miles, which makes our estimated position 36°11′N, 74°12′W, at LAN April 16. The time equivalent of 74°12′ is 0457, which means that the GMT of LAN is 1657.

For P.M. twilight April 16, we again estimate the distance—24 × 3.8 = 91 miles—which puts us at 35°59′N, 73°43′W. At this position, P.M. civil twilight is at 1902 LMT, 2357 GMT. Sunset is at 1835 LMT, 2330 GMT; P.M. nautical twilight is at 1933 LMT, 0028 GMT (April 17).

We continue in the same way, filling in the worksheet for the slow plan. On the last day, April 22, our expected arrival at Gibbs Hill Light is after A.M. twilight and before LAN, so the worksheet is terminated at A.M. twilight.

A Computing Shortcut

If there are periods during the cruise when there are no course changes and the triangle-of-forces computations show a fairly constant expected SMG, the data for the worksheet can be computed for only every third day, and the data for the intervening days can be filled in by interpolation. The middle day on each page of the *Nautical Almanac* should be selected for the computations, because the twilight tables are accurate for that day and only approximate for the other two days on the page. The necessary conditions are satisfied for the middle of our sample cruise, so we could compute the data for the starting date, April 15, then for April 17 and 20, and for the ending days, April 21 and 22, and fill in the worksheet for the other days by interpolation.

Allowing for Uncertain Departure

If the departure date is uncertain, the plan should be started for Day 1, Day 2, etc., and worked as far as possible.

This includes computation of expected average speeds and determination of expected positions at A.M. twilight, LAN, and P.M. twilight for Day 1, Day 2, etc., based on approximate dates. Then the plan can be completed fairly quickly after the departure date has been established.

Moonrise and Moonset

Finding the time of moonrise or moonset requires finding the approximate expected position at the proper time, interpolating for position, getting the LMT, and converting LMT to GMT. This can be done as follows (using April 15 as an example):

1. If it hasn't been done already, mark the planning chart to show the position (on the intended track) at the time of each of the celestial events (A.M. twilight, LAN, and P.M. twilight) each day, using data from the worksheet (Fig. 15).
2. Moonrise on April 15 occurs before our departure, so we will skip moonrise for this date, and start with moonset. Enter the moonset table on the *Nautical Almanac* daily page for April 15 with the approximate expected latitude, 36°45′N, and interpolate to get the LMT of moonset, 1415. Enter this under "Tab." on p. 2 of the worksheet (Fig. 16).
3. In the moonset table, for the nearest tabulated latitude (N35°), find the difference between the listing for this day and the one for the next day: 1516 − 1418 = 58 minutes. Enter Table II, p. xxxii, with 60 minutes and the approximate expected longitude, 75°W, and take out the correction: 12 minutes. Enter this figure on the worksheet under "λ Corr.," marking it +, according to the rule at the bottom of p. xxxii.
4. Find the time equivalent of the longitude (about 75°32′)—0502—and enter it under "Time Equiv. of λ." The longitude where we will be at 1427 LMT (1415 + 12) can be interpolated from the 75°52′ at

DATE	RISING					SETTING			
	LMT		Time Equiv. of λ	GMT		LMT		Time Equiv. of λ	GMT
	Tab.	λ Corr.				Tab.	λ Corr.		
15	—	—	—	—		1415	+12	0502	1929
16	0355	+5	0459	0859		1515	+12	0456	2023
17	0423	+5	0452	0920		1613	+12	0449	2114
18	0446	+5	0446	0937		1708	+12	0442	2202
19	0509	+5	0439	0953		1803	+11	0436	2250
20	0535	+5	0433	1013		1858	+11	0429	2338
21	0603	+6	0426	1035		1953	+11	0423	0027 4/22
22	0635	+6	0420	1101		—	—	—	—

Fig. 16. Worksheet p. 2.

1200 LMT (LAN) and the 75°23′ at 1900 LMT (P.M. civil twilight), or just figured roughly by eye.

5. Add these figures to get the GMT of moonset: 1415 + 12 + 0502 = 1929.

The times for moonrise and moonset for the other days of the cruise are found in a similar manner.

Note that for the moon these computations must be done separately for each day. Linear interpolation should not be used, because the moon's motion is too irregular. Also, do not be confused by the occasional listing (once a month) in the moonrise and moonset tables that shows more than 24 hours. For example, the table shows moonset on April 24, in 52°N latitude, as occurring at 2416. This means that the moon does not set on April 24. It sets that night at 16 minutes after midnight, which makes it April 25.

Preparing the Star Finder and Identifier, H.O. 2102 (Rude's)

Next, we figure the times of rising and setting of the planets each day. The first step is to set up the star finder:

1. Record the GHA♈ each day at 1700 GMT. The time is chosen arbitrarily—1700 is about the time of LAN—but it should be the same for each day. (See Fig. 17.)
2. For each planet except Venus: for the beginning, middle, and last day of the cruise, record on the worksheet the GHA and declination at 1700 GMT. Then record, for the same days at 1700 GMT, the right ascension in degrees (rather than in time units), which equals GHA♈ minus GHA of the planet (or sun or moon), adding 360° as necessary. For a long cruise, this should be done for each week.
3. Record the same information for Venus and the sun, at two-day intervals, and for the moon, every day.
4. Mark with an asterisk each entry on the worksheet for which the GHA of the body is too close to the GHA of the sun for the body to be observed. These figures are, approximately: moon, 25°; Venus, 10°; other planets, 15°.
5. Mark the positions of these bodies, and their dates, on the base of the star finder. Color-code these, if colored pencils are available.

Rising and Setting of Planets

Now that we have set up the star finder, we can use it to find the rising and setting times of the planets each day. This information is needed for getting sights in the middle of the night with the body on the horizon. (See Chapter 10.) The annotated star finder will also be used later to get the predicted altitude and azimuth of each available planet at twilight.

Start with Venus for April 16. (Since Venus rises on April 15 before our departure, we will skip the data for the rising of

CAPE HENRY LIGHT TO GIBBS HILL LIGHT
DATA FOR STAR FINDER
AT 1700 GMT
APRIL 15 – 22, 1985

DATE	GHA♈	SUN			MOON			VENUS			MARS			JUPITER			SATURN			DATE
		GHA	RA	d	GHA	RA	d	GHA	RA	d	GHA	RA	d	GHA	RA	d	GHA	RA	d	
15	99°	75°	24°	N10°	119°	340°	S14°	94°	5°	N8°	49°	50°	N19°	143°	316°	S17°	224°	235°	S17°	15
16	100°	—	—	—	109°	351°	S9°	—	—	—	—	—	—	—	—	—	—	—	—	16
17	101°	75°	26°	N11°	*99°	2°	S3°	96°	5°	N7°	—	—	—	—	—	—	—	—	—	17
18	102°	—	—	—	*89°	13°	N2°	—	—	—	50°	52°	N19°	146°	316°	S17°	227°	235°	S17°	18
19	103°	75°	28°	N11°	*80°	23°	N7°	99°	4°	N7°	—	—	—	—	—	—	—	—	—	19
20	104°	—	—	—	*70°	34°	N11°	—	—	—	—	—	—	—	—	—	—	—	—	20
21	105°	75°	30°	N12°	*60°	45°	N17°	101°	4°	N6°	—	—	—	—	—	—	—	—	—	21
22	106°	75°	31°	N12°	49°	57°	N21°	102°	4°	N6°	51°	55°	N20°	149°	317°	S17°	231°	235°	S17°	22

* TOO CLOSE TO THE SUN FOR OBSERVATION.
DIFFERENCE BETWEEN GHA AND SUN'S GHA MUST BE AT LEAST:
MOON – 25° VENUS – 10° OTHER PLANETS – 15°.

Fig. 17. Worksheet p. 3.

DATE	VENUS								MARS							
	RISING				SETTING				RISING				SETTING			
	LHA♈	GHA♈	GMT	Z_n	LHA♈	GHA♈	GMT	Z_n	LHA♈	GHA♈	GMT	Z_n	LHA♈	GHA♈	GMT	Z_n
15	—	—	—	—	103°	178°	2216	281°	—	—	—	—	—	—	—	—
16	268°	343°	0916	077°	—	—	2204	281°	304°	19°	1136	064°	157°	232°	0152	295°
17	—	—	0905	078°	—	—	2152	280°	—	—	1128	063°	—	—	0145	295°
18	—	—	0853	078°	—	—	2140	280°	—	—	1120	063°	—	—	0137	295°
19	—	—	0842	079°	—	—	2128	280°	—	—	1112	063°	—	—	0130	296°
20	—	—	0833	079°	—	—	2116	279°	—	—	1104	062°	—	—	0123	296°
21	—	—	0822	080°	100°	166°	2104	279°	—	—	1056	062°	—	—	0115	296°
22	268°	333°	0812	080°	—	—	—	—	307°	12°	1048	062°	162°	227°	0108	296°

DATE	JUPITER								SATURN							
	RISING				SETTING				RISING				SETTING			
	LHA♈	GHA♈	GMT	Z_n	LHA♈	GHA♈	GMT	Z_n	LHA♈	GHA♈	GMT	Z_n	LHA♈	GHA♈	GMT	Z_n
15	—	—	—	—	34°	110°	1744	249°	—	—	—	—	—	—	—	—
16	239°	314°	0720	110°	—	—	1733	249°	158°	233°	0156	110°	312°	27°	1208	248°
17	—	—	0709	110°	—	—	1722	249°	—	—	0147	110°	—	—	1158	248°
18	—	—	0658	110°	—	—	1711	249°	—	—	0138	110°	—	—	1148	248°
19	—	—	0647	110°	—	—	1701	249°	—	—	0128	110°	—	—	1138	248°
20	—	—	0637	110°	—	—	1650	249°	—	—	0119	110°	—	—	1128	248°
21	—	—	0626	110°	34°	100°	1640	249°	—	—	0109	110°	—	—	1118	248°
22	239°	304°	0616	110°	—	—	—	—	158°	224°	0056	110°	312°	17°	1108	248°

Fig. 18. Worksheet p. 4.

Venus on that date.) Put the transparent disc for 35°N in place and turn it until the eastern side of the 0° altitude curve falls on the position of Venus for this date, which you marked on the base. (If your star finder is the old model, H.O. 2102-C—whose discs show altitudes down to only 10°—you will need to estimate the 0° altitude curve or measure it with dividers.) Record on p. 4 of the worksheet (Fig. 18), under "Venus—Rising," the LHA♈ (268°) and Zn (077°).

Repeat the process for April 22. Next, noting the approximate expected longitude for each day, as shown on p. 1 of the worksheet, get the GHA ♈ for April 16 and 22 and record it. (GHA♈ = LHA♈ + λW.) Then, for each of the two GHA♈ entries, find the corresponding GMT on the *Nautical Almanac* daily page and record these on the worksheet. Interpolation for GMT can be done by using the table on p. i (the first yellow page) in the *Nautical Almanac*, or more simply by remembering that each 1° equals 4 minutes.

Fill in the GMT and Zn by interpolation for the days between April 16 and April 22. Then, in a similar manner, but using the western side of the 0° altitude curve of the star finder, get the LHA♈ and Zn for the setting of Venus for April 15 and April 21. Get GHA♈ and GMT for these days (as was done for Venus rising). Then fill in the GMT and Zn by interpolation for the days between April 15 and April 21. Using the same methods, complete the worksheet (Fig. 18) for Mars, Jupiter, and Saturn.

The worksheet should be annotated to show those days when a body does not rise or does not set, or does one of these things twice. This does not occur on our sample cruise, but the following example, from 1984, illustrates the situation.

April, 1984
Saturn—Rising

Date	*GMT*	
21	0001	
22	2348	(4/21)

The table, with its annotated entry for April 22, shows that Saturn rises just after midnight beginning April 21 and (having set in the meantime) rises again just before midnight ending April 21.

Sun Positions Marked on Star Finder

We marked the positions of the sun on the star-finder base as a reference, to make it easy to distinguish periods of dark and daylight. They can also be used for checking the times of sunrise and sunset found by interpolating in the *Nautical Almanac* tables.

Keeping the Days Straight

There is a possible source of confusion concerning some of the time listings in the worksheet tables that show both LMT and GMT. Although each line in a table is labelled for one day, it is possible for the LMT and the corresponding GMT for an event to be on different days. For example, on the line labelled April 15 on p. 1 of the worksheet, the LMT of P.M. civil twilight is shown as 1900. However, when this time is converted to GMT, it comes out as 0002—that is, 0002 on April 16. The 0002 entry should be labelled "4/16" to avoid confusion.

Worksheets for the Fast Plan

After the worksheets for the slow plan have been completed, those for the fast plan can be made up without repeating all of the steps. The following data, already prepared for the slow plan, can also be used for the fast plan:

The table showing SMG and TC to steer for each 2° of longitude.

The table showing GHA, RA, and d each day for the sun, moon, and planets (part of worksheet p. 3).

Markings on H.O. 2102 showing positions of the sun, moon, and planets.

LHA♈ and Zn (but not GHA♈ and GMT) at rising and setting of planets (part of worksheet p. 4).

Changing the Plan

If the plan must be revised after it is completed, because the departure date has been changed, the work will not have been wasted. A drastic change—of more than a week—requires recomputing the time of each celestial event (sunrise, sunset, twilight, and LAN), but not the expected positions for the times of these events on the first day of the cruise, second day, etc.

If the departure date has been changed by not more than a week, revision is even easier:

Changes in Plan for Slight Change in Departure Date

Twilight: Use the first-day sheet for the first day, etc.; just change the date on each sheet.

Noonsight: Same as for twilight plans.

Sun-moon fixes: Same as for twilight plans.

Moonrise and moonset: Make a complete revision.

Sights at night—planets on the horizon: Same as for twilight plans if the new departure is within three days of the original one; otherwise revise.

CHAPTER 8

Daily Navigation Guides

Now that the preliminary work is completed, we can make up the final product—the daily navigation guides: a page for each day's A.M. twilight, noonsight, and sun and moon sights; and a page for each day's P.M. twilight sights and sights at night (body on the horizon). There will be one set for the slow plan and one for the fast plan, and they should all be kept in a loose-leaf notebook. Each set of twilight observations is to be recorded (from rough notes made on deck) on the same sheet as the predictions. Data for the other celestial observations should also be in the notebook, arranged by day. The radio and weather logs could be in sections of the same notebook.

Available Bodies and Their Predicted Altitudes and Azimuths

1. For morning sights, list on the daily guide the GMT of nautical twilight (ready to sight), civil twilight (middle of sighting period), and sunrise (end of period). For evening sights, list sunset (ready), civil twilight (middle), and nautical twilight (end).
2. Find LHA♈ at the LMT of civil twilight. Enter the first column of the *Nautical Almanac* daily page (labelled "GMT") with *LMT* (not GMT) of civil twilight and take out *LHA*♈ (not GHA♈) from the column labelled "GHA♈."
3. Enter H.O. 249 Vol. I with the nearest whole degree of predicted latitude and LHA♈, and take out the star name and predicted altitude (rounded to the nearest degree) and azimuth. Mark with a dagger each of the stars so marked in the table. These are the stars that should be observed if time is too short to shoot all seven.
4. Set the star finder (previously marked with the positions of the planets and moon) with the LHA♈ found in 2, above, and take out predicted altitude and azimuth. Check one or two of the star predictions found (per 3, above) in H.O. 249 Vol. I. If this publication is not available, use the star finder for star, planet, and moon predictions. (See Chapter 9 if the Davies *Sight Reduction Tables* are preferred.)
5. List the bodies, with their predicted altitudes and azimuths, in sighting order: dimmest to brightest for A.M., brightest to dimmest for P.M. If brightness is about equal, sight those with easterly azimuths before those with westerly azimuths, A.M. or P.M. Make headings for the columns to be filled in at sighting time with observed data: hs, watch time, and GMT.

Daily Navigation Guide
Tuesday, April 16 (GMT)
Slow Plan

Morning Sights	*GMT*
Nautical twilight (ready to sight)	0925
Civil twilight (middle of sights)	0957
Sunrise (end of sights)	1023

LHA♈ at civil twilight: 279°

		Predicted			*Observed*	
Mag.	*Body*	*h*	*Zn*	*hs*	*Watch*	*GMT*
2.2	Kochab	43°	342°			
2.1	Nunki	28°	175°			
1.9	Alkaid†	37°	309°			
1.3	Deneb†	65°	059°			
1.2	Antares†	21°	210°			
0.9	Altair	58°	144°			
0.4	Saturn	22°	226°			
0.2	Arcturus	31°	272°			
−1.9	Jupiter	26°	140°			
−3.7	Venus	8°	086°			
–	Moon	10°	111°			

Noonsight

GMT of LAN	1657
Predicted altitude	64°06′

Moon

Rises	0859	GMT
Sets	2023	GMT

Fair sun-moon fix available from 1030 to 2015 GMT.

Predicted Time to Begin Sighting

Although the precise time to begin twilight sights depends on a number of factors—including latitude, time of

Daily Navigation Guide
Tuesday/ Wednesday, April 16/17 (GMT)
Slow Plan

Evening Sights	*GMT*
Sunset (ready to sight)	2330
Civil twilight (middle of sights)	2357
Nautical twilight (end of sights)	0028 (4/17)

LHA♈ at civil twilight: 130°

		Predicted			*Observed*	
Mag.	*Body*	*h*	*Zn*	*hs*	*Watch*	*GMT*
−1.6	Sirius	31°	213°			
0.2	Capella	51°	301°			
0.3	Rigel	24°	238°			
1.1	Aldebaran†	33°	267°			
1.3	Regulus†	59°	135°			
1.7	Mars	24°	277°			
2.2	Denebola	43°	104°			
2.2	Kochab†	34°	019°			

Sights at Night—Body on the Horizon (4/17)

Body	*Aspect*	*GMT*	*Zn*
Jupiter	Rising	0709	110°
Mars	Setting	0145	295°
Saturn	Rising	0147	110°

year, and visibility—a close enough approximation for planning can be found fairly easily:

1. For A.M. sights, find the LMT of nautical twilight. For P.M. sights, find the LMT of sunset.
2. Apply to this LMT the time equivalent of longitude (+W, −E), from the first yellow page in the *Nautical Almanac,* to get the GMT to begin the round of sights. This time is actually a little early (it varies from about 10 minutes early in low latitudes to about 30 minutes early in high latitudes, at the summer sol-

Daily Navigation Guide
Friday, April 19 (GMT)
Slow Plan

Morning Sights	*GMT*
Nautical twilight (ready to sight)	0906
Civil twilight (middle of sights)	0937
Sunrise (end of sights)	1003

LHA♈ at civil twilight: 282°

		Predicted			*Observed*	
Mag.	*Body*	*h*	*Zn*	*hs*	*Watch*	*GMT*
2.2	Kochab	41°	342°			
2.1	Nunki†	30°	178°			
1.9	Alkaid†	34°	310°			
1.3	Deneb†	66°	054°			
1.2	Antares	21°	213°			
0.9	Altair	61°	147°			
0.4	Saturn	22°	229°			
0.2	Arcturus	28°	275°			
−1.8	Jupiter	29°	142°			
−3.9	Venus	10°	089°			

Noonsight

GMT of LAN	1636
Predicted altitude	67°11′

Moon

Too close to the sun for observation.

stice), but it allows for the necessary preparations. The exact time to begin sighting will become apparent after the first day's sights are taken.

Repeating Sights

Replication of sights is important to ensure reliable results. For a round of twilight sights, this can be done either by going through the complete list and then repeating it, or by taking two sights on each body in turn. (A duplicate set of "Observed" columns can be provided on the daily navigation guides.) If the second method is used, be sure there is enough time to complete all the planned sights. If you find that the end of twilight is approaching, just get one good sight of each of the bodies remaining on the list. If it is necessary to skip some of the planned shots, be sure to shoot all those marked with a dagger, so that all azimuths will be represented.

Sighting in Cloudy Weather

If the sky is partly cloudy at sighting time, some of the bodies for which the altitude and azimuth were precomputed may be obscured by clouds. The ones that were in the plan and are not obscured can be found easily from the predicted altitude and azimuth, even though the complete pattern of stars cannot be seen. But it may happen that clouds obscure all—or all but one or two—of the stars and planets in the plan. If so, it will be necessary to shoot any bodies that can be seen—even though they were not in the plan—and measure the azimuth as well as the altitude of each one. They can then be identified by the Star Finder and Identifier, and the sights can be worked by H.O. 249 Vol. II or Vol. III (if the declination is not over 30°) or by a backup method. (Identification is not a problem if the Davies *Sight Reduction Tables* are used. See Chapter 9.)

Moon Sights at Twilight

The moon can usually be sighted at evening twilight when it is at least two days past new moon, and at morning twilight when it is at least two days before new moon—that is, when the difference between the moon's GHA and the sun's GHA is at least 25°.

Other Data

Also list on the A.M. daily guides the GMT of LAN, the predicted altitude of the sun at LAN, and the GMT of moonrise and moonset. Add a note to show the hours during which a sun-moon fix is available (if at all), as determined from the star finder and *Nautical Almanac*. On the P.M. daily guides, show (in addition to the twilight data), for planets available during the night, the name, aspect (rising or setting), GMT, and Zn—all from the worksheet.

Planning the Landfall

Plans should include the landfall. The navigator's notebook should have a page for each landfall, showing lights (position, height, visibility, and light characteristics), radio beacons (position, identification, frequency, and range), and commercial stations (frequency and power output).

It is much easier to find this information in one place than to have to search for it on the chart, in the *Light List*, and in *Radio Navigational Aids* at the time of the landfall, when there are other things to do.

Using a Single LOP

When making the landfall, remember that a single line of position can be very useful. If no celestial or electronic fixes have been obtained for several days, and only one LOP is available—from a radio beacon, for instance, or a shot of the sun or another body through a break in the overcast—the boat can be run in to the destination along the LOP.

The procedure is to head the boat a number of miles to one side of the destination (based on the DR) and then turn to a heading along the LOP, toward the destination. The number of miles to offset the track before running in on the LOP is a matter of judgment. Some navigators use 15 percent of the distance run since the last good fix.

Approach from the southwest,
across Challenger Bank.

Radio Beacons

Gibbs Hill Light BDA: 32°15′N, 64°50′W; 295 kHz, 120 miles

St. David's Head BSD: 32°22′N, 64°39′W; 323 kHz

Bermuda Aerobeacon NWU: 32°16′N, 64°52′W, 375 kHz

Weather Broadcasts

Bermuda Radio VRT: 426 kHz, Morse code 0000, 0800, 1200, 1600, 2000 GMT

Bermuda Harbor Radio ZBM: 476 kHz, Morse code 0118, 0518, 1318, 1718, 2018 GMT

Bermuda Harbor Radio ZBM: 2582 kHz, voice (SSB); and 161.95 MHz (Ch. 27), voice (FM) 1235, 2035 GMT, and on receipt

Commercial Broadcast Stations

ZFB	960 kHz	1000W
ZBM-1	1230 kHz	1000W
ZBM-2	1340 kHz	1000W
VSB	1450 kHz	1000W
ZBM-FM	89.1 MHz	1000W
ZFB-FM	94.6 MHz	1000W

Lights

Gibbs Hill: 32°15′N, 64°50′W; 354 ft.; 24 miles; flashing 10 sec.

St. David's Head: 32°22′N, 64°39′W; 220 ft.; 18 miles; occulting 5 sec.

St. Catherine's Point: 32°23′N, 64°41′W; 82 ft.; 10 miles; flashing red

Aerobeacon: 32°22′N, 64°41′W; 141 ft.; 15 miles; alternating WWG 10 sec.

Fig. 19. Bermuda landfall.

CHAPTER 9

The Assumed Altitude Method

This chapter describes recommended planning methods for navigators who use the recently developed tables by Thomas D. Davies—*Star Sight Reduction Tables for 42 Stars: Assumed Altitude Method of Celestial Navigation* and *Sight Reduction Tables for Sun, Moon, and Planets: Assumed Altitude Method of Celestial Navigation.*

A unique feature of the Davies star tables is that they do not require identification of stars by the navigator. This necessitates a slightly different planning routine from the one explained in Chapter 8 for use with H.O. 249 Vol. I.

Davies Star Tables

The star tables are similar in use to those in H.O. 249 Vol. I, but are entered with the nearest whole degree of latitude and observed altitude and with the approximate LHA♈ (rather than with latitude, declination, and LHA♈). The tables give exact LHA♈, star identification, and azimuth (rather than computed altitude and azimuth). They cover latitudes 72°N to 60°S. (H.O. 249 Vol. I covers 89°N to 89°S.)

The main tables show data for 42 stars, ranging in altitude from 15° to 74° (rather than the 7 stars selected for each latitude and LHA♈ combination in H.O. 249 Vol. I). As with H.O. 249 Vol. I, auxiliary tables are included which give GHA♈ for nine years, Polaris correction, and refraction and dip corrections, so the *Nautical Almanac* need not be used. Plotting of LOPs is slightly different from plotting by other methods, but is not complicated.

The main advantage of the Davies star tables is that the stars sighted do not have to be identified in advance. Disadvantages are the need for special treatment of stars near meridian transit, and a more complicated correction for precession and nutation than the one-step correction given in Table 5 of H.O. 249 Vol. I.

Davies Sun, Moon, and Planet Tables

This volume is similar in use to H.O. 249 Vols. II and III, but the tables are entered with the nearest whole degree of latitude, declination, and observed altitude (rather than with latitude, declination, and LHA of the body). The tables give meridian angle and azimuth angle (rather than computed altitude and azimuth angle). They cover latitudes 0° to 60° (compared to 0° to 89° for H.O. 249 Vols. II and III), declinations 0° to 29° or 30° (similar to the coverage in H.O. 249 Vols. II and III), and altitudes 15° to 70° (compared to −6° to 90° in H.O. 249 Vols. II and III).

As with the Davies star tables, special treatment is needed for bodies near meridian transit. Sights of the moon and planets when they are low in the sky cannot be worked by this method, nor can sights at night of bodies on the horizon. LOPs are plotted the same way as for the Davies star tables.

Navigation Planning Using Davies Tables

The Davies star tables could be used to plan twilight sights in the same way H.O. 249 Vol. I is used. It is possible to enter the Davies star tables with latitude and LHA♈ and take out the altitude and azimuth of a group of stars to be sighted at twilight. However, this procedure is not recommended. It would require the scanning of seven pages and deciding which of the 20 or so available stars are the best to sight. Also, it would fail to make use of the advantage the tables provide of not having to identify the stars before sighting.

This is the recommended procedure:

1. Use the star finder to get predicted altitudes and azimuths of available planets and the moon for the predicted time of civil twilight at the predicted position, as described in Chapter 8.
2. Do not do any planning for star sights.
3. Make up the daily navigation guides as described in Chapter 8, but with the star data omitted.
4. When taking the sights, just select bright stars properly spaced in azimuth, and shoot as many as possible, including the planned planets and the moon. Record the approximate azimuth along with the altitude of each body, so that the table listings can be confirmed when the sights are reduced.

CHAPTER 10

Special Problems

Noonsight Projections

The vital need for the noonsight has diminished since accurate time became available at sea in the late eighteenth century, but the custom lives on. Yet it is more than nostalgia that makes the noonsight attractive. Something can always go wrong to cause time errors; also, a noonsight is very easy to work. But, more important, averaging a number of sights taken before, during, and after LAN improves accuracy. Averaging helps for other sights, but is particularly good for the noonsight, which plots as a curve that ascends, peaks, and descends. It is not, as with other sights, an approximation to a straight line, with an unknown slope. Therefore, navigation planning should include the noonsight. The steps are as follows:

1. Find the predicted GMT at LAN (for example, April 15, 1985, at 75°52′W):

LAT of LAN	1200	By definition.
± equation of time	0	*Add* if meridian passage (*Nautical Almanac* daily page) is *after* 1200.
+λW (time equiv.)	0503	First yellow page, *Nautical Almanac.*
GMT of LAN	1703	

2. Find the predicted altitude of the sun at LAN, using the appropriate formula:

Dec. contrary to lat.: $h = 90° - L - d$
Dec. same, < lat.: $h = 90° + d - L$
Dec. same, > lat.: $h = 90° + L - d$

High-Altitude Sun Sights

The altitude prediction will facilitate finding the sun if the sky is partly cloudy at LAN. Also, it will show whether the sun will be at too high an altitude to take a normal noonsight—that is, by taking repeated sights as the sun's altitude increases, peaks, then decreases, and using the peak altitude as the LAN sight (or drawing a smooth curve through the points and using the top of the curve as the LAN sight). If the sun's altitude will be above 85° at LAN, it is better to take a sight before LAN and another after LAN, and plot a running fix. In plotting any sights with Ho close to 90°, a special technique is required because the circle of position has such a small radius that it does not approximate a straight line. An ordinary LOP, if not corrected, would be inaccurate.

One alternative for dealing with this problem is, first, to find the geographical position (GP) for the first sight: latitude equals declination; longitude, if W, equals GHA. (For E longitude, use 360° minus GHA.) Advance this GP for the boat's run between sights. Draw a circle of position with radius equal to the coaltitude (90° minus Ho). Then do the same for the second sight (except for advancing), and the point where the circles intersect is the running fix. There will be two such points, but one can be eliminated by comparison with the DR position or by noting the sun's azimuth.

The second choice for handling high-altitude sights is to use the Table of Offsets, p. xx, H.O. 229, to get a correction, and then plot as usual.

Sun-and-Moon Fixes during the Day

The navigator who is planning an offshore cruise should include preparations for sun-and-moon fixes during the day for all those days when the moon is favorably placed. The moon should be shot (in addition to the sun) whenever it is visible during the day, regardless of the angle of cut that the moon's LOP makes with the sun's LOP, because moon sights can give a useful check on the sun sights, even if the LOPs are nearly parallel. And if the LOPs cross at more than about 30°, there is the advantage of having a sun-moon fix during daylight.

This kind of fix is entirely reliable, if it is done without error. But because the built-in checking system of a round of star sights (having five or six LOPs to check against one another) is not available, and because moon sights require a few more computations—with more opportunities for error—the navigator should take at least five sights on each body, use special care in working the sights (repeating each step), and compare the fix with the DR position—then rework anything that doesn't look right. The whole process should be repeated several hours later if the moon is still favorably placed.

Celestial Sights at Night

One problem navigators have had for many years is the need to get celestial sights in the middle of the night when the horizon is not visible. Sometimes around full moon part of the horizon may be lighted sufficiently by moonlight to make it possible to get a few sights, but these occasions are fairly rare, and it may be a false horizon that is seen.

Some books describe a procedure for lifeboat or other emergency use, when there is no sextant available. It requires sighting the sun or moon by eye alone or with binoculars when the limb (upper or lower) just touches the horizon at rising or setting, taking "hs" as 0°00′ making the usual corrections, and computing an LOP. (Use eye protection when sighting the sun.)

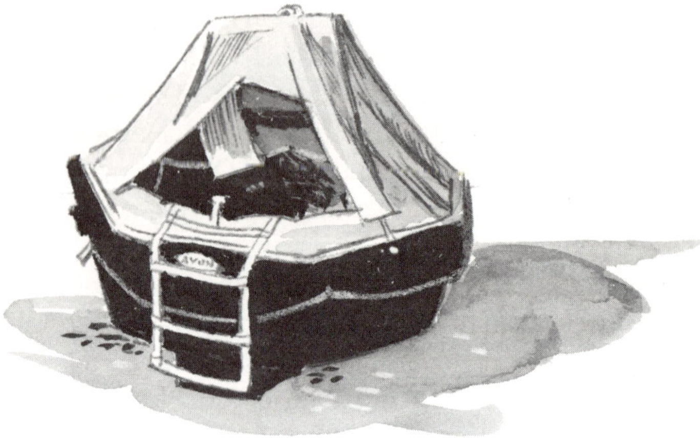

Recommended Method. This will work just as well when there is no emergency, and can be used in clear weather in the middle of the night with planets and bright stars, in addition to the moon. Even though the horizon is not distinct, a setting star or planet marks its arrival at the horizon by disappearing—and the reverse for rising bodies. When the moon is just starting to set, the horizon will "take a bite" out of the lower limb.

There is one thing to be careful of: this method does not require that the horizon be seen, but it does require a clear sky at the horizon. The navigator should guard against a low-lying cloud bank (with a clear sky above) that could give the same effect as the horizon in blanking out a star or planet. Binoculars and a long viewing period can help; in any case, caution is advised. The recommended procedure is as follows.

Planning before the Cruise. For each day of the cruise, find the azimuth and the GMT at rising and setting of the moon and planets. Ignore the times for rising and setting during the day (using the GMT of sunrise and sunset as a guide), and list the remainder. (See Chapters 7 and 8.)

Choice of Bodies. Try to select a group of bodies that will be on the horizon at about the same time, preferably some with azimuths a little north and some a little south of east or west to get a good cut of LOPs for a running fix.

Taking Sights at Night. When taking a sight on a rising planet, stand up high on deck and scan the horizon at the precomputed time and around the precomputed azimuth. When the planet has just risen, drop your head or move to a lower position. (This procedure is difficult, but worth trying. Setting bodies are easier.) Record the GMT and height of eye the moment the planet disappears. Compute the LOP.

When taking a sight on a setting planet, sit in the cockpit and watch the planet as it sets. When it disappears below the horizon, move to a higher position, then drop to the height where the planet just disappears. Record the GMT and height of eye and compute the LOP.

Moon sights are similar to planet sights, but easier.

Star sights can be treated the same as planet sights if the stars are bright enough.

Working the Sight. Getting the LOP is straightforward; however, as for all low-altitude sights, *Nautical Almanac* Table A4, "Additional Corrections," should be used, besides the usual corrections.

Also, when the corrections are applied to 0°00′ "hs," Ho will often be negative, and Hc may or may not be negative, so a little care is needed to avoid mistakes with the plus and minus signs.

H.O. 229 has an explanation in the front of the book for its somewhat complicated procedure for handling negative altitudes. H.O. 249, in Vols. II and III, very helpfully tabulates negative altitudes directly, down to −6°, and is the recommended method.

A short review is in order for those who have not worked with negative numbers recently. For normal sights—when Ho and Hc are both positive—the "toward" or "away" decision for plotting the LOP with respect to the azimuth can be made by remembering:

Computed greater, *away*.
Observed greater, *toward*.

For example:

Hc	35°17′	
Ho	35°05′	
	12	*away* from Zn (because Hc is greater than Ho).

The same memory guide (or a similar one) can be used for sights on the horizon, but now we have to deal with negative numbers in two ways; first, in applying the d correction to the tabulated altitude. Note that we are algebraically adding d to the tabulated altitude:

	Case 1	*Case 2*	*Case 3*	*Case 4*	*Case 5*
Tabulated h	00°39′	−00°14′	−00°47′	00°17′	−00°20′
d correction	−16′	−12′	+25′	−28′	+25′
Hc	00°23′	−00°26′	−00°22′	−00°11′	00°05′

Secondly, negative numbers can come up when Hc is compared with Ho to get the intercept and decide whether it is toward the Zn or away:

	Case 1	*Case 2*	*Case 3*	*Case 4*
Hc	00°07′	−00°16′	−00°10′	−00°06′
Ho	−00°12′	−00°13′	−00°15′	00°09′
Intercept	19 (not 5)	3	5	15 (not 3)
Direction	away	toward (because −13′ is greater than −16′)	away (because −10′ is greater than −15′)	toward

Note that we are taking the algebraic *difference* between Hc and Ho, not algebraically adding them. This means that we change the sign of the second number (Ho) and algebraically add. The sign of the intercept (+ or −) is ignored when the work is finished.

APPENDIX A

Sight-Reduction Form for H. O. 249 or H. O. 229

Date:__________ Assumed lat.:________*

DR pos.:______________ Ht. of eye:______ft.

at________ GMT

Body							
GMT							
hs							
IC							
dip							
R/LL/UL							
Ho							
GHA (h)	v						
SHA☆							
GHA (m/s)							
v corr. (moon, planets)							
GHA (total)							
−λW or +λE							
LHA		*	*	*	*	*	*
Dec.	*d*						
d corr. (not for stars)							
Dec. (total)		*	*	*	*	*	*
Altitude	d						
d corr.							
Hc							
Ho							
Intercept							
Z							
Zn							

* Table entry

Notes for Sight-Reduction Form

Form Entry	H.O. 249 Vol. I Stars	H.O. 249 Vol. II or III; H.O. 229			
		Sun	Moon	Planet	Star
GHA (h)	GHA♈ (hour)	← GHA of the body (hour) →			
v	–	–	*Nautical Almanac* white pages		–
SHA ☆	–	–	–	–	*Naut. Alm.* white pages
GHA (m/s)	← *Nautical Almanac* yellow pages: 1st, 2nd, or 3rd column →				
v corr.	–	–	*Nautical Almanac* yellow pages		–
GHA (total)	GHA♈	GHA sun	GHA moon	GHA planet	GHA star
−λW / +λE	← Use a figure close to the DRλ that will result in a whole degree of LHA →				
LHA	LHA♈	LHA sun	LHA moon	LHA planet	LHA star
Dec.	–	← *Nautical Almanac* white pages; Dec. (hr.) →			
d	–	← *Nautical Almanac* white pages →			–
d corr	–	← *Nautical Almanac* yellow pages: dec. inc. for min. and sec. (show + or −) →			–
Dec. (total)	–	← Total declination →			
Altitude	–	← Called "Hc" in H.O. 249 Vols. II and III, and H.O. 229 →			
d	–	← H.O. 249 Vols. II and III, or H.O. 229: "d", main table (show + or −) →			
d corr.	–	H.O. 249 Vols. II and III, Table III, or H.O. 229 Interpolation Table (show + or −) →			
Hc	Tabulated	← "Alt." ± "d corr." →			
Ho	← From the upper section of the form →				
Intercept	← Difference between Hc and Ho ("computed greater away") →				
Z	–	← H.O. 249 Vols. II and III, or H.O. 229 main table →			
Zn	Main table Also use Table 5, p. 322	← Computed from Z, following the rules on each page of H.O. 249 Vols. II and III, or H.O. 229 →			

APPENDIX B

Sources of Publications

Government Publications

Pilot Charts

Atlas of Pilot Charts—South Atlantic Ocean
Atlas of Pilot Charts—Central American Waters and South Atlantic
Atlas of Pilot Charts—South Pacific Ocean
Atlas of Pilot Charts—Northern North Atlantic
Atlas of Pilot Charts—Indian Ocean
Pilot Chart of the North Atlantic Ocean
January–March
April–June
July–September
October–December
Pilot Chart of the North Pacific Ocean
January–March
April–June
July–September
October–December

Defense Mapping Agency
Hydrographic/Topographic Center
Washington, D.C. 20315

Ocean Passages for the World

Available from agents for the sale of British Admiralty charts

Nautical Almanac

U. S. Government Printing Office
Washington, D. C. 20402

Also available from:

Her Majesty's Stationery Office
49 High Holborn
London W. C. 1, England

Air Almanac
Same as for the *Nautical Almanac*

Sight Reduction Tables for Air Navigation—
Pub. No. 249 (AP3270 in the U. K.)
Vol. I—Selected Stars
Vol. II—Latitudes 0°–40°, Declinations 0°–29°
Vol. III—Latitudes 40°–89°, Declinations 0°–29°
Defense Mapping Agency or U. S. Government Printing Office

Sight Reduction Tables for Marine Navigation—
Pub. No. 229 (NP401 in the U. K.)
Vol. 1—Latitudes 0°–15°
Vol. 2—Latitudes 15°–30°
Vol. 3—Latitudes 30°–45°
Vol. 4—Latitudes 45°–60°
Vol. 5—Latitudes 60°–75°
Vol. 6—Latitudes 75°–90°
Defense Mapping Agency or U. S. Government Printing Office

Selected Worldwide Marine Weather Broadcasts
U. S. Government Printing Office

Radio Navigational Aids
Pub. 117A—Atlantic and Mediterranean Area
Pub. 117B—Pacific and Indian Oceans Area
Defense Mapping Agency

Radio Aids to Marine Navigation
Atlantic and Great Lakes
Pacific
Printing and Publishing Supply and Services
Hull, Quebec, Canada K1A OS9

United States Coast Guard Light Lists
Vol. I—Atlantic Coast (St. Croix River, Maine to Little River, South Carolina)
Vol. II—Atlantic and Gulf Coast (Little River, South Carolina to Rio Grande, Texas and Greater Antilles)
Vol. III—Pacific Coast and Pacific Islands . . .
Vol. IV—Great Lakes, U.S. and Canada
Vol. V—Mississippi River System
U. S. Government Printing Office

List of Lights Including Fog Signals
Pub. 110—Greenland, the East Coasts of North and South America, and the West Indies
Pub. 111—West Coasts of North and South America, Australia, Tasmania, New Zealand, and the islands of the North and South Pacific Oceans
Pub. 112—Western Pacific and Indian Oceans
Pub. 113—West Coasts of Europe and Africa, the Mediterranean Sea, Black Sea, and Azovskoye More
Pub. 114—British Isles, English Channel, and North Sea
Pub. 115—Norway, Iceland, and Arctic Ocean
Pub. 116—Baltic Sea, Kattegat, Belts and Sound, and Gulf of Bothnia
Defense Mapping Agency
American Practical Navigator
Vols. I and II
Defense Mapping Agency or U. S. Government Printing Office

Non-Government Publications

Bayless, Allan E. *Compact Sight Reduction Table (Modified H.O. 211, Ageton's Table)*. Centreville, Md.: Cornell Maritime Press, 1980.

Davies, Thomas D. *Sight Reduction Tables for Sun, Moon, and Planets: Assumed Altitude Method of Celestial Navigation.* Centreville, Md.: Cornell Maritime Press, 1982.

———. *Star Sight Reduction Tables for 42 Stars: Assumed Altitude Method of Celestial Navigation.* Centreville, Md.:Cornell Maritime Press, 1980.

Howell, Susan P. *Practical Celestial Navigation.* Hebron, Conn.: Susan P. Howell, 1979.

Letcher, John S., Jr. *Self-Contained Celestial Navigation with H.O. 208.* Camden, Me.: International Marine Publishing Co., 1977.

Milligan, John E. *Celestial Navigation by H.O. 249.* Centreville, Md.: Cornell Maritime Press, 1974.

Moody, Alton B. *Navigation Afloat.* New York: Van Nostrand Reinhold Co., 1980.

Noer, H. Rolf. *Navigator's Pocket Calculator Handbook.* Centreville, Md.: Cornell Maritime Press, 1983.

Schlereth, Hewitt. *Commonsense Celestial Navigation.* Chicago: Henry Regnery Co., 1975.

Wright, Frances W. *Celestial Navigation.* 2d ed. Centreville, Md.: Cornell Maritime Press, 1982.

———. *Particularized Navigation: How to Prevent Navigational Emergencies.* Cambridge, Md.: Cornell Maritime Press, 1973.

APPENDIX C

Sight-Reduction Methods Compared

H.O. 249 (AP3270), Vol. I and Vol. II or III. The first choice for small boats. Vol. I is excellent for planning and reducing star sights, and for identification of stars that are among the seven listed for a given latitude and LHA ♈. Vol. II (lat. 0°–40°) or III (lat. 40°–89°) is good for reducing sights of the sun, moon, and planets, and can be used for stars with declinations 0° to 29°. It is also very good for reducing sights of bodies on the horizon.

H.O. 229 (NP401), Vols. 1–6. The second choice for small boats. Slightly more accurate than H.O. 249, but bulkier, more expensive, and more time-consuming. Covers all latitudes and declinations. Star and planet identification is not easy. (Use H.O. 2102 instead.)

H.O. 214 (HD486), Vols. I–IX. Similar to H.O. 229, but a little less convenient. Obsolescent but still useful.

H.O. 211. Out of print by the government, but sometimes included in navigation texts. Also, an improved version by Bayless, with only 9 pages of tables (compared to 36 in the original) and simplified instructions, is available. Similar to H.O. 208, but can be used for sight reduction with the DR as assumed position, and for computing great-circle solutions. The Bayless version is first choice for lifeboat and backup use.

H.O. 208. Very compact and inexpensive. Very time-consuming. Not to be considered for star and planet identification if another method is available. Good for lifeboat use and as a backup to other methods. Out of print (by the government) but sometimes included in navigation texts.

H.O. 218 (AP618), in 14 Volumes. Forerunner of H.O. 249. Not as good as H.O. 249 Vol. I for planning and star identification; otherwise, fast and easy to use. Out of print, but can be used if found.

Davies Star Tables. These tables have a unique feature: they do not require identification of stars before sighting. They cover the brightest 42 stars, of which about half are visible at any one time. The basic method cannot be used for stars near meridian transit, and the book provides a graph for solution of these sights by reduction to the meridian. Corrections for precession and nutation—for years other than the ones for which the tables were compiled—are somewhat involved. (See Chapter 9 for further comments.)

Davies Sun, Moon, and Planet Tables. Similar in purpose to H.O. 249 Vols. II and III, but with table entries and plotting of LOPs related to those of the Davies *Star Tables,* rather than to H.O. 249 and similar methods. The tables are limited to bodies with altitudes 15° to 70°, compared to −6° to 90° in H.O. 249 Vols. II and III. They can be used for any bodies, including stars, within these limits. As with the Davies *Star Tables,* sights of stars that are near meridian transit are worked by reduction to the meridian, using a graph provided in the book. (See Chapter 9 for further comments.)

Specialized Navigation Calculators. These have some of the *Nautical Almanac* tables incorporated in their memory. They are fast and precise, but expensive, and should be protected from salt air. They are excellent for planning and identification, figuring great-circle courses, and other navigation problems, and can be used with the DR as assumed position. Do not depend on one without a backup method, in case of failure. Take spare batteries or power packs.

Generalized Programmable Calculators. When programmed (see Chapter 5), these can be used with the *Nautical Almanac* or *Air Almanac* to give precise results. Not as good as H.O. 249 Vols. I, II, and III for planning or for ordinary sight reduction, but better for computing great-circle courses and for sight reduction using the DR as assumed position. Also better for identification of an unknown star or planet that was sighted. Do not depend on one without a backup method. Take spare batteries or power packs.

Engineering Calculators. Too slow for regular use in sight reduction, but good for backup and for adding and subtracting numbers when tabular methods are used. Also good for interpolation in the *Nautical Almanac,* calculating speeds, and general computations. Inexpensive; at least one should be on board. Should be protected from salt air. Take spare batteries.

Sine-Cosine Formula. A last-resort emergency method. Good for lifeboats. Cost and bulk are close to zero. Must be used with tables of sines and cosines—preferably, also, with log-sine and log-cosine tables.

APPENDIX D

Checklists

Checklist 1—Navigation Equipment

Sextant, with spare mirrors
3 timepieces (recommended: 2 quartz digital, 1 spring-wound)
Nautical Almanac, Air Almanac, or *Reed's Almanac*
H.O. 249 (AP3270), Vol. I, or Davies *Star Tables*
H.O. 249 (AP3270), Vol. II or III, or H.O. 229 (NP401), or
    Davies *Sun, Moon, and Planet Tables*
Charts for departure and landfall; plotting charts for open
    water
Plotting tools
    Paper (plain and graph), pencils, erasers
    Dividers
    Protractor-plotter
    Small plotting board
Star Finder and Identifier, H.O. 2102 (Rude's)
Prepared navigation plans
Log book
Selected Worldwide Marine Weather Broadcasts
Radio Navigational Aids
Light List
List of Lights
Navigation text
Tables from Bowditch, *American Practical Navigator*, Pub.
    No. 9
Tide tables and current tables
Pocket calculator, protected from corrosion
Spare dry cells—alkaline, protected from corrosion—and, if
    needed, power packs
Small AM radio for picking up static for thunderstorm
    warnings

Other portable electronic equipment, if used (RDF, etc.)
Binoculars
Hand-bearing compass
Equipment manuals
Seasickness medicine

Checklist 2—Survival Navigation Equipment

EPIRB emergency transmitter
Plastic sextant
Nautical Almanac
H.O. 211, H.O. 208, or trigonometric tables and formulas
Plotting charts
Paper, pencils, eraser
Dividers
Plotter-protractor
Small radio receiver and spare batteries
Flashlights and spare batteries
Small compass

Checklist 3—Navigator's Preparations

Height of eye of navigator and assistants—on deck (forward and aft) and in the cockpit.
Boat's length at the waterline.
Electronic equipment—check operation.
Electronic equipment—protect with plastic covers and silica gel.
Sextants—clean mirrors and shades and adjust if necessary; check operation by working sights taken from a known position (with artificial horizon, if necessary).
Compass—check by swinging.
RDF—check error on various headings.

APPENDIX E

Formulas

Sight-Reduction Formulas

$$\text{Sin Hc} = (\sin L \sin d) + (\cos L \cos d \cos LHA)$$

$$\text{Cos Z} = \frac{\sin d - (\sin L \sin Hc)}{\cos L \cos Hc}$$

If cos Z is negative, look up arc cos Z as for a positive cos, then subtract this from 180° to get Z. Find Zn from Z:

$$\text{If LHA} \geq 180°, \text{Zn} = \text{Z}$$
$$\text{If LHA} < 180°, \text{Zn} = 360° - \text{Z}$$

Rules:

N lat. +, S lat. −
N dec. +, S dec. −
The sin of a negative angle is −. } Most calculators apply
The cos of a negative angle is +. } these rules automatically.

	LHA (Θ)	*Conversion*	*Cos*
Conversion and sign needed for LHA if LHA > 90°	90° to 180°	180° − Θ	−
	180° to 270°	Θ − 180°	−
	270° to 360°	360° − Θ	+

These formulas cover all latitudes and declinations. They can be worked either with an engineering calculator or by hand with tables of sines and cosines—preferably also log-sines and log-consines.

Noonsight Formulas

If d is contrary to L:	L = 90° − Ho − d
If d is the same name and < L:	L = 90° − Ho + d
If d is the same name and > L:	L = Ho + d − 90°

Equation of Time

To convert from apparent to mean time:

If meridian passage is *after* 12 hr., *add* Eq. T.
If meridian passage is *before* 12 hr., *subtract* Eq. T.

Interpolation with a Calculator

Example: HP-33C or HP 33-E
(Other RPN calculators are similar.)

Given: For April 16, 1985, the *Nautical Almanac* lists the LMT of A.M. civil twilight as follows:

Lat.	*LMT*
N40°	0451
N35°	0500

Find: LMT of civil twilight at 36°23′N.

Solution (using linear regression):

f CLEAR REG	
51	(minutes for 0451)
ENTER	
40	(40°)
Σ +	
60	(minutes for 0500)
ENTER	
35	(35°)
Σ +	
36.38	(36°23′ as decimal degrees)
f $\hat{y}$	
Read 57.5160	(or, rounded, 58 minutes)

Answer: 0458, the LMT of civil twilight at 36°23′N.

Finding the True Wind

This is not a difficult problem. It is necessary, however, to keep "to" and "from" wind directions straight—a wind is spoken of as being *from* a certain direction, but the arrow for the vector diagram is drawn *to* the reciprocal of that direction.

The problem could be worked on an engineering calculator, but is more easily visualized if it is solved graphically, using a small plotting board or just paper, pencil, protractor, and ruler.

The apparent-wind vector is the resultant of the "boat's-wind" vector (the reciprocal of the boat's heading) and the true-wind vector:

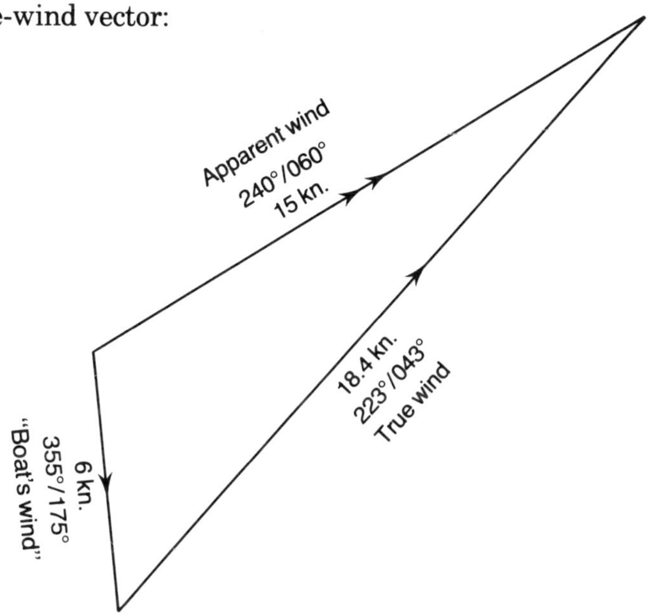

Fig. 20. Finding true wind.

Example

Given: Boat's heading is 355°, speed 6 knots. Apparent wind is 15 knots from 240°.

Find: The true wind.

Solution:

1. Using a convenient scale, draw the vector for the boat's wind in the direction of the reciprocal of the heading—175°, 6 units long.

2. From the *tail* of the boat's-wind vector, draw the apparent-wind vector toward 060° (the reciprocal of 240°), 15 units long.
3. Measure the angle (043°) and the length (18.4 units) of the true-wind vector from the head of the boat's-wind vector to the head of the apparent-wind vector. The true wind, then, is 18.4 knots from 223° (the reciprocal of 043°).

Storm Avoidance

The complete solution to a storm-avoidance problem is somewhat complicated. Fortunately, the most important information—the course to steer—can be found in a few minutes with an engineering calculator. Sometimes this is all that is needed. In any case, the process will usually have to be repeated as the storm changes course—a good reason for finding a quick solution. The first thing to do is draw a simple diagram to visualize the problem.

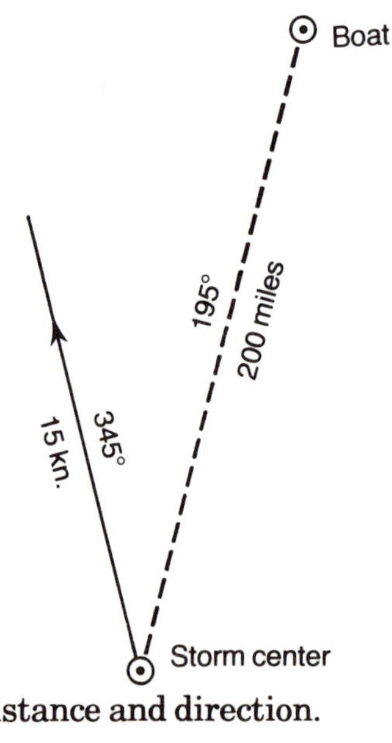

Fig. 21. Storm's distance and direction.

Example

Given: A storm center bears 195°, 200 miles from the boat. The storm is moving toward 345° at 15 knots. The boat can make 6 knots in the expected direction (more or less northeasterly, as can be inferred from the diagram).

Find: The best course to avoid the storm.

Solution:

$$\text{Course difference} = \text{arc cos}\frac{\text{boat speed}}{\text{storm speed}}$$
$$= \text{arc cos}\frac{6}{15}$$
$$= 66°$$
$$\text{Best course} = 345° + 66°(-360°) = 051°$$

Note that we measured the 66° course difference to the right (away from the storm), rather than to the left (toward the storm).

Now that the most urgent problem has been solved, and the boat is on its new course away from the storm, the time-consuming remainder of the problem can be worked. A vector diagram gives the easiest solution here.

Solving the Vector Diagram

1. Draw the storm vector CA, 345°, 15 knots, using a convenient scale.
2. Draw the boat's vector CB, 051°, 6 knots. (The 051° direction was found in the calculator solution, preceding.)
3. Draw BA to complete the first triangle. Measure BA, 321°, 13.7 knots. (Check to see that the angle at B is 90°. If it is not, check the computations.)
4. In the second triangle, draw CE, 195°, 200 miles, using another convenient scale.
5. Extend BC.
6. Draw ED parallel to BA and perpendicular to BC extended. This locates point D.
7. Measure CD (to the same scale that was used for CE). This gives 162 miles, the storm's closest approach.

8. Measure ED (to the same scale as for CD and CE). This gives 118 miles of relative movement. Then, 118 ÷ 13.7 (the speed of relative movement) = 8.6, or 8 hours 36 minutes until the storm's closest approach.

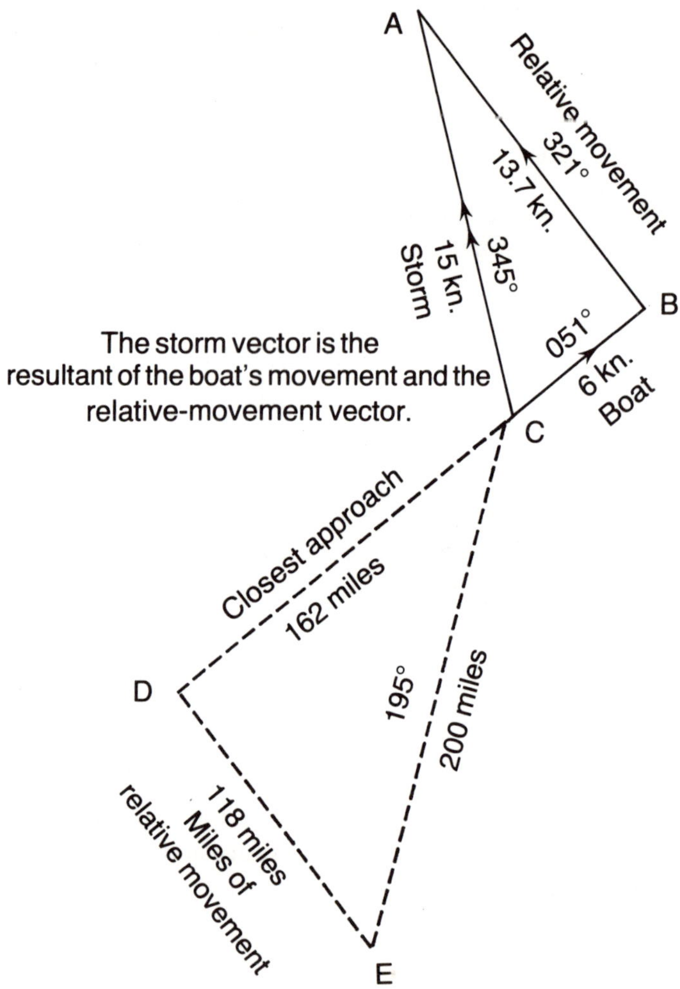

Fig. 22. Storm avoidance—vector diagram.

APPENDIX F

Glossary

♈	Aries. The point on the celestial sphere from which the hour angle of stars and other bodies is measured. The point at which SHA = 0°.
θ	General symbol for an angle.
$\geq$	Equal to or greater than.
$\leq$	Equal to or less than.
$>$	Greater than.
$<$	Less than.
λ	Longitude.
*a*L	Assumed latitude.
$a\lambda$	Assumed longitude.
altitude	The angle, at the observer, from the horizon to a celestial body. (See hs, Ho, and Hc.).
AP	Assumed position. Usually chosen at a whole degree of latitude, and with a longitude that will make the LHA a whole degree.
d	Correction to declination for minutes and seconds of time (in the *NauticalAlmanac*).
d	Correction to Hc for minutes of declination (in H.O. 249, H.O. 229, Davies *Sight Reduction Tables*, etc.).
dec. or d	Declination. Analogous to latitude. The angle from the plane of the celestial equator, toward the north or south pole, to a celestial body.
dip	Correction to hs for height of eye. Always minus; equal (in minutes) to $0.97\sqrt{h}$ (where h = height of eye above sea level, in feet).
DR	Dead reckoning. Determination of position from an earlier known position and distance and direction of movement.

Eq. T.	Equation of time. The difference, in minutes of time, between the position of the mean sun (a fictional concept), on which time is based, and the apparent (real) sun, whose motion is less regular. To convert from apparent to mean time, add the Eq. T. if meridian passage is after 1200; otherwise, subtract.
GHA	Greenwich hour angle. The angle from the meridian of Greenwich, measured westward to a celestial body (0° to 360°). GHA☆ =GHA♈ + SHA☆
GMT	The local mean time at 0° longitude (the meridian of Greenwich).
GP	Geographical position. The point on the earth directly below a celestial body; the point at which Ho = 90°.
h	Same as altitude.
Hc	Computed altitude of a celestial body, at a particular time, at the AP selected for a tabular method, such as H.O. 249.
HE	Height of eye of an observer above sea level.
Ho	Observed altitude. The result of applying all the necessary corrections to hs.
HP	Horizontal parallax (of the moon). The parallax at 0° altitude.
hs	The angular altitude of a celestial body above the horizon, measured by a sextant, before corrections are made to get Ho.
IC	Index correction, which must be made to a sextant reading to get the correct figure.
Int.	Intercept. The difference between Ho and Hc, stated in nautical miles. Remember "computed greater away"—that is, if Hc is greater than Ho, the LOP is plotted away from the azimuth, a distance equal to the intercept.

L	Latitude.
LAN	Local apparent noon. The time when the sun is at its highest altitude at a particular place and the local apparent time (LAT) is 1200.
LAT	Local apparent time. The angle measured westward from the observer's meridian to the apparent (real) sun and converted to time units.
LHA	Local hour angle. Measured from the observer's meridian, westward, to a celestial body (0° to 360°). GHA − λW (or + λE) = LHA.
LL	Lower limb. The lower edge (of the sun or moon). *Nautical Almanac* corrections for LL observations are the total of refraction, parallax, and semidiameter corrections.
LMT	Local mean time. Equal to GMT minus (for west λ) or plus (for east λ) the time equivalent of λ.
LOP	Line of position.
LZT	Local zone time. Differs from GMT by one hour for each 15° of longitude. LZT is similar to LMT, but LZT changes in one-hour steps, and LMT changes continuously.
N.A.	The *Nautical Almanac.*
N.M.	Nautical miles. One nautical mile equals 6080 feet, or one minute of latitude.
P	Parallax. The error caused by an observer's not being at the center of the earth. The amount is large for the moon, small for the sun, and negligible for other bodies.
R	Refraction. The error caused by the bending of light rays (from a celestial body) by the atmosphere. The correction is always minus.
RA	Right ascension. RA = GHA♈ − GHA of the body.

SD	Semidiameter. The radius (of the sun or moon). The SD correction is + for lower-limb observations, − for upper-limb observations.
SHA	Sidereal hour angle. Measured from ♈, westward, to a star (0° to 360°).
SMG	Speed made good (along the desired track).
UL	Upper limb. The upper edge (of the sun or moon). *Nautical Almanac* corrections of UL observations are the total of refraction, parallax, and semidiameter corrections.
v	Correction to GHA for minutes and seconds of time (in the *Nautical Almanac*). For planets and the moon only.
Z	Azimuth angle. Measured from the north or south, toward east or west (0° to 180°).
Zn	Azimuth. Measured from the north toward east (0° to 360°).

APPENDIX G

Excerpts

Nautical Almanac

Pub. No. 249 (H.O. 249), Vol. I

1985 APRIL 13, 14, 15 (SAT., SUN., MON.)

G.M.T. (UT) d h	ARIES G.H.A.	VENUS −3.7 G.H.A.	VENUS Dec.	MARS +1.7 G.H.A.	MARS Dec.	JUPITER −1.7 G.H.A.	JUPITER Dec.	SATURN +0.4 G.H.A.	SATURN Dec.
	° ′	° ′	° ′	° ′	° ′	° ′	° ′	° ′	° ′
13 00	201 07.4	195 25.7	N 9 10.6	153 03.0	N18 17.3	245 40.9	S17 24.4	325 47.3	S17 13.2
01	216 09.9	210 29.0	09.7	168 03.7	17.8	260 43.0	24.3	340 49.9	13.2
02	231 12.4	225 32.3	08.7	183 04.4	18.3	275 45.1	24.2	355 52.5	13.1
03	246 14.8	240 35.6	07.8	198 05.1	18.8	290 47.2	24.1	10 55.1	13.1
04	261 17.3	255 38.9	06.8	213 05.7	19.3	305 49.3	24.0	25 57.7	13.0
05	276 19.8	270 42.3	05.9	228 06.4	19.7	320 51.4	23.9	41 00.3	13.0
06	291 22.2	285 45.6	N 9 05.0	243 07.1	N18 20.2	335 53.5	S17 23.8	56 02.9	S17 13.0
07	306 24.7	300 48.9	04.0	258 07.8	20.7	350 55.6	23.7	71 05.5	12.9
S 08	321 27.2	315 52.2	03.1	273 08.5	21.2	5 57.7	23.6	86 08.1	12.9
A 09	336 29.6	330 55.5	02.1	288 09.1	21.7	20 59.8	23.5	101 10.7	12.9
T 10	351 32.1	345 58.8	01.2	303 09.8	22.2	36 01.9	23.4	116 13.3	12.8
U 11	6 34.6	1 02.1	9 00.2	318 10.5	22.7	51 04.0	23.3	131 15.9	12.8
R 12	21 37.0	16 05.3	N 8 59.3	333 11.2	N18 23.2	66 06.1	S17 23.2	146 18.5	S17 12.8
D 13	36 39.5	31 08.6	58.4	348 11.9	23.7	81 08.2	23.1	161 21.2	12.7
A 14	51 41.9	46 11.9	57.4	3 12.5	24.2	96 10.3	23.0	176 23.8	12.7
Y 15	66 44.4	61 15.2	56.5	18 13.2	24.6	111 12.4	22.9	191 26.4	12.7
16	81 46.9	76 18.5	55.6	33 13.9	25.1	126 14.5	22.8	206 29.0	12.6
17	96 49.3	91 21.7	54.6	48 14.6	25.6	141 16.6	22.7	221 31.6	12.6
18	111 51.8	106 25.0	N 8 53.7	63 15.2	N18 26.1	156 18.7	S17 22.6	236 34.2	S17 12.6
19	126 54.3	121 28.3	52.8	78 15.9	26.6	171 20.8	22.5	251 36.8	12.5
20	141 56.7	136 31.5	51.8	93 16.6	27.1	186 22.9	22.4	266 39.4	12.5
21	156 59.2	151 34.8	50.9	108 17.3	27.6	201 25.0	22.3	281 42.0	12.4
22	172 01.7	166 38.0	50.0	123 18.0	28.1	216 27.1	22.3	296 44.6	12.4
23	187 04.1	181 41.3	49.0	138 18.6	28.5	231 29.2	22.2	311 47.2	12.4
14 00	202 06.6	196 44.5	N 8 48.1	153 19.3	N18 29.0	246 31.3	S17 22.1	326 49.8	S17 12.3
01	217 09.1	211 47.8	47.2	168 20.0	29.5	261 33.4	22.0	341 52.4	12.3
02	232 11.5	226 51.0	46.2	183 20.7	30.0	276 35.5	21.9	356 55.0	12.3
03	247 14.0	241 54.2	45.3	198 21.4	30.5	291 37.6	21.8	11 57.6	12.2
04	262 16.4	256 57.5	44.4	213 22.0	31.0	306 39.7	21.7	27 00.2	12.2
05	277 18.9	272 00.7	43.5	228 22.7	31.5	321 41.8	21.6	42 02.8	12.2
06	292 21.4	287 03.9	N 8 42.5	243 23.4	N18 31.9	336 43.9	S17 21.5	57 05.5	S17 12.1
07	307 23.8	302 07.1	41.6	258 24.1	32.4	351 46.0	21.4	72 08.1	12.1
08	322 26.3	317 10.3	40.7	273 24.7	32.9	6 48.1	21.3	87 10.7	12.0
S 09	337 28.8	332 13.6	39.8	288 25.4	33.4	21 50.2	21.2	102 13.3	12.0
U 10	352 31.2	347 16.8	38.9	303 26.1	33.9	36 52.3	21.1	117 15.9	12.0
N 11	7 33.7	2 20.0	37.9	318 26.8	34.3	51 54.4	21.0	132 18.5	11.9
D 12	22 36.2	17 23.2	N 8 37.0	333 27.4	N18 34.8	66 56.5	S17 20.9	147 21.1	S17 11.9
A 13	37 38.6	32 26.4	36.1	348 28.1	35.3	81 58.6	20.8	162 23.7	11.9
Y 14	52 41.1	47 29.6	35.2	3 28.8	35.8	97 00.7	20.7	177 26.3	11.8
15	67 43.5	62 32.7	34.3	18 29.5	36.3	112 02.8	20.6	192 28.9	11.8
16	82 46.0	77 35.9	33.4	33 30.2	36.8	127 04.9	20.5	207 31.5	11.8
17	97 48.5	92 39.1	32.4	48 30.8	37.2	142 07.0	20.5	222 34.1	11.7
18	112 50.9	107 42.3	N 8 31.5	63 31.5	N18 37.7	157 09.1	S17 20.4	237 36.7	S17 11.7
19	127 53.4	122 45.5	30.6	78 32.2	38.2	172 11.2	20.3	252 39.3	11.7
20	142 55.9	137 48.6	29.7	93 32.9	38.7	187 13.3	20.2	267 42.0	11.6
21	157 58.3	152 51.8	28.8	108 33.5	39.2	202 15.4	20.1	282 44.6	11.6
22	173 00.8	167 55.0	27.9	123 34.2	39.6	217 17.6	20.0	297 47.2	11.5
23	188 03.3	182 58.1	27.0	138 34.9	40.1	232 19.7	19.9	312 49.8	11.5
15 00	203 05.7	198 01.3	N 8 26.1	153 35.6	N18 40.6	247 21.8	S17 19.8	327 52.4	S17 11.5
01	218 08.2	213 04.4	25.2	168 36.2	41.1	262 23.9	19.7	342 55.0	11.4
02	233 10.7	228 07.6	24.3	183 36.9	41.5	277 26.0	19.6	357 57.6	11.4
03	248 13.1	243 10.7	23.4	198 37.6	42.0	292 28.1	19.5	13 00.2	11.4
04	263 15.6	258 13.9	22.5	213 38.3	42.5	307 30.2	19.4	28 02.8	11.3
05	278 18.0	273 17.0	21.6	228 38.9	43.0	322 32.3	19.3	43 05.4	11.3
06	293 20.5	288 20.2	N 8 20.7	243 39.6	N18 43.5	337 34.4	S17 19.2	58 08.0	S17 11.3
07	308 23.0	303 23.3	19.8	258 40.3	43.9	352 36.5	19.1	73 10.7	11.2
08	323 25.4	318 26.4	18.9	273 41.0	44.4	7 38.6	19.1	88 13.3	11.2
M 09	338 27.9	333 29.5	18.0	288 41.6	44.9	22 40.7	19.0	103 15.9	11.1
O 10	353 30.4	348 32.7	17.1	303 42.3	45.4	37 42.8	18.9	118 18.5	11.1
N 11	8 32.8	3 35.8	16.2	318 43.0	45.8	52 44.9	18.8	133 21.1	11.1
D 12	23 35.3	18 38.9	N 8 15.3	333 43.7	N18 46.3	67 47.1	S17 18.7	148 23.7	S17 11.0
A 13	38 37.8	33 42.0	14.4	348 44.3	46.8	82 49.2	18.6	163 26.3	11.0
Y 14	53 40.2	48 45.1	13.5	3 45.0	47.3	97 51.3	18.5	178 28.9	11.0
15	68 42.7	63 48.2	12.6	18 45.7	47.7	112 53.4	18.4	193 31.5	10.9
16	83 45.2	78 51.3	11.7	33 46.4	48.2	127 55.5	18.3	208 34.2	10.9
17	98 47.6	93 54.4	10.9	48 47.0	48.7	142 57.6	18.2	223 36.8	10.8
18	113 50.1	108 57.5	N 8 10.0	63 47.7	N18 49.2	157 59.7	S17 18.1	238 39.4	S17 10.8
19	128 52.5	124 00.6	09.1	78 48.4	49.6	173 01.8	18.0	253 42.0	10.8
20	143 55.0	139 03.7	08.2	93 49.1	50.1	188 03.9	17.9	268 44.6	10.7
21	158 57.5	154 06.7	07.3	108 49.7	50.6	203 06.0	17.8	283 47.2	10.7
22	173 59.9	169 09.8	06.4	123 50.4	51.0	218 08.2	17.8	298 49.8	10.7
23	189 02.4	184 12.9	05.6	138 51.1	51.5	233 10.3	17.7	313 52.4	10.6
Mer. Pass.	h m 10 29.8	v 3.2	d 0.9	v 0.7	d 0.5	v 2.1	d 0.1	v 2.6	d 0.0

STARS Name	S.H.A.	Dec.
	° ′	° ′
Acamar	315 35.0	S40 21.9
Achernar	335 43.1	S57 18.8
Acrux	173 33.1	S63 01.1
Adhara	255 29.5	S28 57.2
Aldebaran	291 14.4	N16 28.8
Alioth	166 38.7	N56 02.4
Alkaid	153 15.2	N49 23.1
Al Na'ir	28 10.8	S47 02.0
Alnilam	276 08.4	S 1 12.7
Alphard	218 17.1	S 8 35.7
Alphecca	126 28.9	N26 45.6
Alpheratz	358 06.2	N29 00.3
Altair	62 29.2	N 8 49.4
Ankaa	353 37.1	S42 23.2
Antares	112 52.5	S26 24.1
Arcturus	146 15.0	N19 15.4
Atria	108 13.4	S69 00.0
Avior	234 26.9	S59 27.9
Bellatrix	278 55.3	N 6 20.2
Betelgeuse	271 24.8	N 7 24.3
Canopus	264 05.9	S52 41.4
Capella	281 06.6	N45 59.2
Deneb	49 46.4	N45 13.2
Denebola	182 55.3	N14 39.3
Diphda	349 17.7	S18 04.2
Dubhe	194 17.1	N61 50.1
Elnath	278 40.1	N28 35.9
Eltanin	90 55.9	N51 29.0
Enif	34 08.4	N 9 48.1
Fomalhaut	15 47.8	S29 42.1
Gacrux	172 24.7	S57 01.9
Gienah	176 14.2	S17 27.7
Hadar	149 18.2	S60 18.2
Hamal	328 25.5	N23 23.5
Kaus Aust.	84 12.2	S34 23.6
Kochab	137 17.6	N74 12.8
Markab	14 00.1	N15 07.3
Menkar	314 37.9	N 4 01.9
Menkent	148 32.7	S36 17.9
Miaplacidus	221 44.2	S69 39.6
Mirfak	309 11.8	N49 48.6
Nunki	76 24.9	S26 19.1
Peacock	53 53.0	S56 46.9
Pollux	243 54.0	N28 03.9
Procyon	245 22.2	N 5 15.8
Rasalhague	96 26.3	N12 33.9
Regulus	208 06.2	N12 02.4
Rigel	281 32.9	S 8 13.2
Rigil Kent.	140 20.8	S60 46.4
Sabik	102 37.1	S15 42.6
Schedar	350 06.0	N56 27.2
Shaula	96 51.0	S37 05.7
Sirius	258 52.8	S16 41.8
Spica	158 53.7	S11 05.2
Suhail	223 08.3	S43 22.5
Vega	80 53.4	N38 45.8
Zuben'ubi	137 29.0	S15 59.0

	S.H.A.	Mer. Pass.
	° ′	h m
Venus	354 37.9	10 51
Mars	311 12.7	13 46
Jupiter	44 24.7	7 33
Saturn	124 43.2	2 12

1985 APRIL 13, 14, 15 (SAT., SUN., MON.)

G.M.T. (UT) d h	SUN G.H.A. ° ′	SUN Dec. ° ′	MOON G.H.A. ° ′	v ′	Dec. ° ′	d ′	H.P. ′
13 00	179 50.2	N 8 56.8	254 56.4	09.3	S24 28.6	06.8	56.3
01	194 50.4	57.7	269 24.7	09.3	24 21.8	07.0	56.3
02	209 50.5	58.6	283 53.0	09.4	24 14.8	07.0	56.3
03	224 50.7	8 59.5	298 21.4	09.6	24 07.8	07.2	56.2
04	239 50.8	9 00.4	312 50.0	09.6	24 00.6	07.3	56.2
05	254 51.0	01.4	327 18.6	09.7	23 53.3	07.4	56.2
06	269 51.2	N 9 02.3	341 47.3	09.8	S23 45.9	07.5	56.1
07	284 51.3	03.2	356 16.1	10.0	23 38.4	07.6	56.1
08	299 51.5	04.1	10 45.1	10.0	23 30.8	07.7	56.1
09	314 51.6	. . 05.0	25 14.1	10.1	23 23.1	07.9	56.1
10	329 51.8	05.9	39 43.2	10.3	23 15.2	07.9	56.0
11	344 52.0	06.8	54 12.5	10.3	23 07.3	08.1	56.0
12	359 52.1	N 9 07.7	68 41.8	10.4	S22 59.2	08.2	56.0
13	14 52.3	08.6	83 11.2	10.5	22 51.0	08.3	55.9
14	29 52.4	09.5	97 40.7	10.7	22 42.7	08.3	55.9
15	44 52.6	. . 10.4	112 10.4	10.7	22 34.4	08.5	55.9
16	59 52.7	11.3	126 40.1	10.8	22 25.9	08.6	55.9
17	74 52.9	12.2	141 09.9	10.9	22 17.3	08.7	55.8
18	89 53.1	N 9 13.1	155 39.8	11.1	S22 08.6	08.8	55.8
19	104 53.2	14.0	170 09.9	11.1	21 59.8	08.9	55.8
20	119 53.4	14.9	184 40.0	11.2	21 50.9	08.9	55.7
21	134 53.5	. . 15.8	199 10.2	11.3	21 42.0	09.1	55.7
22	149 53.7	16.7	213 40.5	11.4	21 32.9	09.2	55.7
23	164 53.8	17.6	228 10.9	11.5	21 23.7	09.2	55.7
14 00	179 54.0	N 9 18.5	242 41.4	11.6	S21 14.5	09.4	55.6
01	194 54.2	19.4	257 12.0	11.7	21 05.1	09.4	55.6
02	209 54.3	20.3	271 42.7	11.8	20 55.7	09.6	55.6
03	224 54.5	. . 21.2	286 13.5	11.9	20 46.1	09.6	55.6
04	239 54.6	22.1	300 44.4	12.0	20 36.5	09.7	55.5
05	254 54.8	23.0	315 15.4	12.1	20 26.8	09.8	55.5
06	269 54.9	N 9 23.9	329 46.5	12.1	S20 17.0	09.8	55.5
07	284 55.1	24.8	344 17.6	12.3	20 07.2	10.0	55.5
08	299 55.3	25.7	358 48.9	12.3	19 57.2	10.0	55.4
09	314 55.4	. . 26.6	13 20.2	12.5	19 47.2	10.2	55.4
10	329 55.6	27.5	27 51.7	12.5	19 37.0	10.2	55.4
11	344 55.7	28.4	42 23.2	12.6	19 26.8	10.3	55.4
12	359 55.9	N 9 29.3	56 54.8	12.7	S19 16.5	10.3	55.3
13	14 56.0	30.2	71 26.5	12.8	19 06.2	10.5	55.3
14	29 56.2	31.1	85 58.3	12.9	18 55.7	10.5	55.3
15	44 56.3	. . 32.0	100 30.2	13.0	18 45.2	10.6	55.3
16	59 56.5	32.9	115 02.2	13.1	18 34.6	10.6	55.2
17	74 56.6	33.8	129 34.3	13.1	18 24.0	10.8	55.2
18	89 56.8	N 9 34.7	144 06.4	13.2	S18 13.2	10.8	55.2
19	104 57.0	35.6	158 38.6	13.4	18 02.4	10.8	55.2
20	119 57.1	36.5	173 11.0	13.4	17 51.6	11.0	55.1
21	134 57.3	. . 37.4	187 43.4	13.4	17 40.6	11.0	55.1
22	149 57.4	38.3	202 15.8	13.6	17 29.6	11.1	55.1
23	164 57.6	39.2	216 48.4	13.7	17 18.5	11.1	55.1
15 00	179 57.7	N 9 40.1	231 21.1	13.7	S17 07.4	11.2	55.1
01	194 57.9	41.0	245 53.8	13.8	16 56.2	11.3	55.0
02	209 58.0	41.9	260 26.6	13.9	16 44.9	11.3	55.0
03	224 58.2	. . 42.8	274 59.5	13.9	16 33.6	11.4	55.0
04	239 58.3	43.7	289 32.4	14.1	16 22.2	11.5	55.0
05	254 58.5	44.6	304 05.5	14.1	16 10.7	11.5	55.0
06	269 58.6	N 9 45.5	318 38.6	14.2	S15 59.2	11.6	54.9
07	284 58.8	46.4	333 11.8	14.2	15 47.6	11.6	54.9
08	299 58.9	47.3	347 45.0	14.4	15 36.0	11.7	54.9
09	314 59.1	. . 48.2	2 18.4	14.4	15 24.3	11.8	54.9
10	329 59.2	49.1	16 51.8	14.5	15 12.5	11.8	54.9
11	344 59.4	49.9	31 25.3	14.5	15 00.7	11.8	54.8
12	359 59.5	N 9 50.8	45 58.8	14.7	S14 48.9	11.9	54.8
13	14 59.7	51.7	60 32.5	14.6	14 37.0	12.0	54.8
14	29 59.8	52.6	75 06.1	14.8	14 25.0	12.0	54.8
15	45 00.0	. . 53.5	89 39.9	14.8	14 13.0	12.1	54.8
16	60 00.1	54.4	104 13.7	14.9	14 00.9	12.1	54.7
17	75 00.3	55.3	118 47.6	15.0	13 48.8	12.2	54.7
18	90 00.4	N 9 56.2	133 21.6	15.0	S13 36.6	12.2	54.7
19	105 00.6	57.1	147 55.6	15.1	13 24.4	12.2	54.7
20	120 00.7	58.0	162 29.7	15.2	13 12.2	12.3	54.7
21	135 00.9	. . 58.9	177 03.9	15.2	12 59.9	12.4	54.7
22	150 01.0	9 59.7	191 38.1	15.2	12 47.5	12.4	54.6
23	165 01.2	10 00.6	206 12.3	15.4	12 35.1	12.4	54.6
	S.D. 16.0	d 0.9	S.D.	15.2	15.1		14.9

Lat.	Twilight Naut. h m	Twilight Civil h m	Sunrise h m	Moonrise 13 h m	14 h m	15 h m	16 h m
N 72	////	02 08	03 45	▬	▬	06 52	05 58
N 70	////	02 42	04 01	▬	▬	06 15	05 40
68	00 58	03 06	04 14	▬	06 27	05 48	05 25
66	01 50	03 24	04 24	06 42	05 48	05 27	05 13
64	02 21	03 39	04 33	05 34	05 20	05 11	05 03
62	02 43	03 51	04 40	04 58	04 58	04 57	04 55
60	03 00	04 01	04 47	04 32	04 41	04 45	04 47
N 58	03 15	04 10	04 52	04 12	04 26	04 35	04 41
56	03 27	04 17	04 57	03 55	04 14	04 26	04 35
54	03 37	04 24	05 02	03 41	04 03	04 18	04 30
52	03 46	04 30	05 06	03 28	03 53	04 11	04 25
50	03 54	04 36	05 10	03 17	03 44	04 05	04 21
45	04 10	04 47	05 17	02 54	03 25	03 51	04 12
N 40	04 23	04 56	05 24	02 35	03 10	03 39	04 04
35	04 33	05 04	05 30	02 19	02 57	03 29	03 57
30	04 41	05 10	05 35	02 06	02 46	03 21	03 51
20	04 54	05 21	05 43	01 42	02 26	03 05	03 41
N 10	05 04	05 29	05 50	01 22	02 09	02 52	03 32
0	05 12	05 36	05 57	01 03	01 53	02 40	03 23
S 10	05 18	05 42	06 04	00 44	01 37	02 27	03 14
20	05 23	05 48	06 11	00 24	01 20	02 14	03 05
30	05 26	05 54	06 18	00 00	01 00	01 59	02 55
35	05 28	05 57	06 23	24 49	00 49	01 50	02 49
40	05 29	06 00	06 28	24 35	00 35	01 39	02 42
45	05 30	06 04	06 34	24 19	00 19	01 27	02 33
S 50	05 30	06 08	06 41	24 00	00 00	01 13	02 24
52	05 30	06 09	06 44	23 50	25 06	01 06	02 19
54	05 30	06 11	06 47	23 40	24 58	00 58	02 14
56	05 30	06 13	06 51	23 28	24 49	00 49	02 08
58	05 30	06 15	06 55	23 14	24 40	00 40	02 02
S 60	05 29	06 18	07 00	22 57	24 28	00 28	01 55

Lat.	Sunset h m	Twilight Civil h m	Twilight Naut. h m	Moonset 13 h m	14 h m	15 h m	16 h m
N 72	20 19	22 00	////	▬	▬	11 11	13 34
N 70	20 02	21 24	////	▬	▬	11 47	13 50
68	19 49	20 59	23 19	▬	09 58	12 12	14 03
66	19 39	20 40	22 17	08 00	10 37	12 31	14 13
64	19 30	20 25	21 44	09 07	11 04	12 47	14 21
62	19 22	20 12	21 21	09 42	11 24	12 59	14 29
60	19 15	20 02	21 03	10 08	11 41	13 10	14 35
N 58	19 10	19 53	20 48	10 28	11 55	13 19	14 40
56	19 05	19 45	20 36	10 44	12 07	13 28	14 45
54	19 00	19 38	20 25	10 58	12 17	13 35	14 49
52	18 56	19 32	20 16	11 10	12 26	13 41	14 53
50	18 52	19 26	20 08	11 21	12 35	13 47	14 57
45	18 44	19 15	19 52	11 43	12 52	13 59	15 05
N 40	18 37	19 05	19 39	12 01	13 06	14 10	15 11
35	18 32	18 58	19 29	12 16	13 18	14 18	15 16
30	18 27	18 51	19 20	12 29	13 29	14 26	15 21
20	18 18	18 40	19 07	12 51	13 46	14 39	15 29
N 10	18 10	18 32	18 57	13 10	14 02	14 50	15 36
0	18 04	18 25	18 49	13 28	14 16	15 01	15 43
S 10	17 57	18 18	18 43	13 45	14 30	15 11	15 50
20	17 50	18 12	18 38	14 04	14 46	15 23	15 56
30	17 42	18 06	18 34	14 25	15 03	15 35	16 04
35	17 37	18 03	18 32	14 38	15 13	15 43	16 09
40	17 32	17 59	18 31	14 52	15 24	15 51	16 14
45	17 26	17 56	18 30	15 09	15 37	16 00	16 20
S 50	17 19	17 52	18 29	15 30	15 53	16 12	16 27
52	17 16	17 50	18 29	15 40	16 01	16 17	16 30
54	17 12	17 48	18 29	15 51	16 09	16 23	16 34
56	17 09	17 46	18 29	16 03	16 19	16 29	16 37
58	17 04	17 44	18 30	16 18	16 29	16 37	16 42
S 60	16 59	17 42	18 30	16 35	16 41	16 45	16 47

Day	SUN Eqn. of Time 00 h m s	SUN Eqn. of Time 12 h m s	Mer. Pass. h m	MOON Mer. Pass. Upper h m	Mer. Pass. Lower h m	Age d	Phase
13	00 39	00 32	12 01	07 15	19 41	23	
14	00 24	00 17	12 00	08 05	20 28	24	
15	00 09	00 02	12 00	08 50	21 12	25	

1985 APRIL 16, 17, 18 (TUES., WED., THURS.)

	G.M.T. (UT) d h	ARIES G.H.A. ° ′	VENUS −3.8 G.H.A. ° ′	VENUS Dec. ° ′	MARS +1.7 G.H.A. ° ′	MARS Dec. ° ′	JUPITER −1.8 G.H.A. ° ′	JUPITER Dec. ° ′	SATURN +0.4 G.H.A. ° ′	SATURN Dec. ° ′	STARS Name	S.H.A. ° ′	Dec. ° ′
	16 00	204 04.9	199 15.9	N 8 04.7	153 51.8	N18 52.0	248 12.4	S17 17.6	328 55.0	S17 10.6	Acamar	315 35.0	S40 21.9
	01	219 07.3	214 19.0	03.8	168 52.4	52.5	263 14.5	17.5	343 57.7	10.6	Achernar	335 43.1	S57 18.8
	02	234 09.8	229 22.1	02.9	183 53.1	52.9	278 16.6	17.4	359 00.3	10.5	Acrux	173 33.1	S63 01.2
	03	249 12.3	244 25.1	· · 02.1	198 53.8	· · 53.4	293 18.7	· · 17.3	14 02.9	· · 10.5	Adhara	255 29.6	S28 57.2
	04	264 14.7	259 28.2	01.2	213 54.4	53.9	308 20.8	17.2	29 05.5	10.4	Aldebaran	291 14.4	N16 28.8
	05	279 17.2	274 31.2	8 00.3	228 55.1	54.3	323 22.9	17.1	44 08.1	10.4			
	06	294 19.6	289 34.3	N 7 59.4	243 55.8	N18 54.8	338 25.1	S17 17.0	59 10.7	S17 10.4	Alioth	166 38.7	N56 02.4
	07	309 22.1	304 37.3	58.6	258 56.5	55.3	353 27.2	16.9	74 13.3	10.3	Alkaid	153 15.1	N49 23.1
T	08	324 24.6	319 40.3	57.7	273 57.1	55.7	8 29.3	16.8	89 15.9	10.3	Al Na'ir	28 10.8	S47 02.0
U	09	339 27.0	334 43.4	· · 56.8	288 57.8	· · 56.2	23 31.4	· · 16.7	104 18.6	· · 10.3	Alnilam	276 08.4	S 1 12.7
E	10	354 29.5	349 46.4	56.0	303 58.5	56.7	38 33.5	16.6	119 21.2	10.2	Alphard	218 17.1	S 8 35.7
S	11	9 32.0	4 49.4	55.1	318 59.2	57.1	53 35.6	16.6	134 23.8	10.2			
D	12	24 34.4	19 52.5	N 7 54.3	333 59.8	N18 57.6	68 37.7	S17 16.5	149 26.4	S17 10.1	Alphecca	126 28.9	N26 45.6
A	13	39 36.9	34 55.5	53.4	349 00.5	58.1	83 39.9	16.4	164 29.0	10.1	Alpheratz	358 06.2	N29 00.3
Y	14	54 39.4	49 58.5	52.5	4 01.2	58.6	98 42.0	16.3	179 31.6	10.1	Altair	62 29.2	N 8 49.4
	15	69 41.8	65 01.5	· · 51.7	19 01.8	· · 59.0	113 44.1	· · 16.2	194 34.2	· · 10.0	Ankaa	353 37.1	S42 23.2
	16	84 44.3	80 04.5	50.8	34 02.5	59.5	128 46.2	16.1	209 36.8	10.0	Antares	112 52.5	S26 24.1
	17	99 46.8	95 07.5	50.0	49 03.2	18 59.9	143 48.3	16.0	224 39.5	10.0			
	18	114 49.2	110 10.5	N 7 49.1	64 03.9	N19 00.4	158 50.4	S17 15.9	239 42.1	S17 09.9	Arcturus	146 15.0	N19 15.4
	19	129 51.7	125 13.5	48.2	79 04.5	00.9	173 52.5	15.8	254 44.7	09.9	Atria	108 13.4	S69 00.0
	20	144 54.1	140 16.5	47.4	94 05.2	01.3	188 54.7	15.7	269 47.3	09.8	Avior	234 26.9	S59 27.9
	21	159 56.6	155 19.5	· · 46.5	109 05.9	· · 01.8	203 56.8	· · 15.6	284 49.9	· · 09.8	Bellatrix	278 55.3	N 6 20.2
	22	174 59.1	170 22.5	45.7	124 06.5	02.3	218 58.9	15.6	299 52.5	09.8	Betelgeuse	271 24.8	N 7 24.3
	23	190 01.5	185 25.4	44.8	139 07.2	02.7	234 01.0	15.5	314 55.1	09.7			
	17 00	205 04.0	200 28.4	N 7 44.0	154 07.9	N19 03.2	249 03.1	S17 15.4	329 57.8	S17 09.7	Canopus	264 05.9	S52 41.4
	01	220 06.5	215 31.4	43.2	169 08.6	03.7	264 05.3	15.3	345 00.4	09.7	Capella	281 06.6	N45 59.2
	02	235 08.9	230 34.4	42.3	184 09.2	04.1	279 07.4	15.2	0 03.0	09.6	Deneb	49 46.4	N45 13.2
	03	250 11.4	245 37.3	· · 41.5	199 09.9	· · 04.6	294 09.5	· · 15.1	15 05.6	· · 09.6	Denebola	182 55.3	N14 39.3
	04	265 13.9	260 40.3	40.6	214 10.6	05.1	309 11.6	15.0	30 08.2	09.6	Diphda	349 17.7	S18 04.2
	05	280 16.3	275 43.2	39.8	229 11.2	05.5	324 13.7	14.9	45 10.8	09.5			
	06	295 18.8	290 46.2	N 7 38.9	244 11.9	N19 06.0	339 15.8	S17 14.8	60 13.4	S17 09.5	Dubhe	194 17.1	N61 50.1
W	07	310 21.3	305 49.1	38.1	259 12.6	06.4	354 18.0	14.7	75 16.1	09.4	Elnath	278 40.1	N28 35.9
E	08	325 23.7	320 52.1	37.3	274 13,3	06.9	9 20.1	14.6	90 18.7	09.4	Eltanin	90 55.9	N51 29.0
D	09	340 26.2	335 55.0	· · 36.4	289 13.9	· · 07.4	24 22.2	· · 14.6	105 21.3	· · 09.4	Enif	34 08.4	N 9 48.1
N	10	355 28.6	350 58.0	35.6	304 14.6	07.8	39 24.3	14.5	120 23.9	09.3	Fomalhaut	15 47.8	S29 42.1
E	11	10 31.1	6 00.9	34.8	319 15.3	08.3	54 26.4	14.4	135 26.5	09.3			
S	12	25 33.6	21 03.8	N 7 33.9	334 15.9	N19 08.7	69 28.6	S17 14.3	150 29.1	S17 09.3	Gacrux	172 24.7	S57 01.9
D	13	40 36.0	36 06.7	33.1	349 16.6	09.2	84 30.7	14.2	165 31.8	09.2	Gienah	176 14.2	S17 27.7
A	14	55 38.5	51 09.7	32.3	4 17.3	09.7	99 32.8	14.1	180 34.4	09.2	Hadar	149 18.2	S60 18.2
Y	15	70 41.0	66 12.6	· · 31.5	19 18.0	· · 10.1	114 34.9	· · 14.0	195 37.0	· · 09.1	Hamal	328 25.5	N23 23.5
	16	85 43.4	81 15.5	30.6	34 18.6	10.6	129 37.0	13.9	210 39.6	09.1	Kaus Aust.	84 12.2	S34 23.6
	17	100 45.9	96 18.4	29.8	49 19.3	11.0	144 39.2	13.8	225 42.2	09.1			
	18	115 48.4	111 21.3	N 7 29.0	64 20.0	N19 11.5	159 41.3	S17 13.7	240 44.8	S17 09.0	Kochab	137 17.5	N74 12.8
	19	130 50.8	126 24.2	28.2	79 20.6	12.0	174 43.4	13.7	255 47.4	09.0	Markab	14 00.1	N15 07.3
	20	145 53.3	141 27.1	27.4	94 21.3	12.4	189 45.5	13.6	270 50.1	08.9	Menkar	314 37.9	N 4 01.9
	21	160 55.7	156 30.0	· · 26.5	109 22.0	· · 12.9	204 47.7	· · 13.5	285 52.7	· · 08.9	Menkent	148 32.7	S36 18.0
	22	175 58.2	171 32.9	25.7	124 22.6	13.3	219 49.8	13.4	300 55.3	08.9	Miaplacidus	221 44.3	S69 39.6
	23	191 00.7	186 35.8	24.9	139 23.3	13.8	234 51.9	13.3	315 57.9	08.8			
	18 00	206 03.1	201 38.7	N 7 24.1	154 24.0	N19 14.2	249 54.0	S17 13.2	331 00.5	S17 08.8	Mirfak	309 11.8	N49 48.6
	01	221 05.6	216 41.5	23.3	169 24.6	14.7	264 56.1	13.1	346 03.1	08.8	Nunki	76 24.9	S26 19.1
	02	236 08.1	231 44.4	22.5	184 25.3	15.2	279 58.3	13.0	1 05.8	08.7	Peacock	53 52.9	S56 46.9
	03	251 10.5	246 47.3	· · 21.7	199 26.0	· · 15.6	295 00.4	· · 12.9	16 08.4	· · 08.7	Pollux	243 54.0	N28 03.9
	04	266 13.0	261 50.2	20.9	214 26.7	16.1	310 02.5	12.8	31 11.0	08.6	Procyon	245 22.3	N 5 15.8
	05	281 15.5	276 53.0	20.1	229 27.3	16.5	325 04.6	12.8	46 13.6	08.6			
	06	296 17.9	291 55.9	N 7 19.2	244 28.0	N19 17.0	340 06.8	S17 12.7	61 16.2	S17 08.6	Rasalhague	96 26.2	N12 33.9
	07	311 20.4	306 58.7	18.4	259 28.7	17.4	355 08.9	12.6	76 18.9	08.5	Regulus	208 06.2	N12 02.4
T	08	326 22.9	322 01.6	17.6	274 29.3	17.9	10 11.0	12.5	91 21.5	08.5	Rigel	281 33.0	S 8 13.1
H	09	341 25.3	337 04.4	· · 16.8	289 30.0	· · 18.3	25 13.1	· · 12.4	106 24.1	· · 08.5	Rigil Kent.	140 20.8	S60 46.5
U	10	356 27.8	352 07.3	16.0	304 30.7	18.8	40 15.3	12.3	121 26.7	08.4	Sabik	102 37.0	S15 42.6
R	11	11 30.2	7 10.1	15.3	319 31.3	19.2	55 17.4	12.2	136 29.3	08.4			
S	12	26 32.7	22 13.0	N 7 14.5	334 32.0	N19 19.7	70 19.5	S17 12.1	151 31.9	S17 08.3	Schedar	350 06.0	N56 27.2
D	13	41 35.2	37 15.8	13.7	349 32.7	20.1	85 21.6	12.0	166 34.6	08.3	Shaula	96 50.9	S37 05.7
A	14	56 37.6	52 18.6	12.9	4 33.3	20.6	100 23.8	12.0	181 37.2	08.3	Sirius	258 52.8	S16 41.8
Y	15	71 40.1	67 21.4	· · 12.1	19 34.0	· · 21.0	115 25.9	· · 11.9	196 39.8	· · 08.2	Spica	158 53.7	S11 05.2
	16	86 42.6	82 24.3	11.3	34 34.7	21.5	130 28.0	11.8	211 42.4	08.2	Suhail	223 08.3	S43 22.5
	17	101 45.0	97 27.1	10.5	49 35.3	22.0	145 30.2	11.7	226 45.0	08.2			
	18	116 47.5	112 29.9	N 7 09.7	64 36.0	N19 22.4	160 32.3	S17 11.6	241 47.7	S17 08.1	Vega	80 53.4	N38 45.8
	19	131 50.0	127 32.7	08.9	79 36.7	22.9	175 34.4	11.5	256 50.3	08.1	Zuben'ubi	137 29.0	S15 59.0
	20	146 52.4	142 35.5	08.2	94 37.3	23.3	190 36.5	11.4	271 52.9	08.0			
	21	161 54.9	157 38.3	· · 07.4	109 38.0	· · 23.8	205 38.7	· · 11.3	286 55.5	· · 08.0			
	22	176 57.4	172 41.1	06.6	124 38.7	24.2	220 40.8	11.3	301 58.1	08.0			
	23	191 59.8	187 43.9	05.8	139 39.3	24.7	235 42.9	11.2	317 00.8	07.9			
	Mer. Pass.	h m 10 18.0	v 2.9	d 0.8	v 0.7	d 0.5	v 2.1	d 0.1	v 2.6	d 0.0			

	S.H.A. ° ′	Mer. Pass. h m
Venus	355 24.4	10 36
Mars	309 03.9	13 43
Jupiter	43 59.1	7 23
Saturn	124 53.8	2 00

1985 APRIL 16, 17, 18 (TUES., WED., THURS.)

G.M.T. (UT) d h	SUN G.H.A. ° ′	SUN Dec. ° ′	MOON G.H.A. ° ′	v ′	Dec. ° ′	d ′	H.P. ′
16 00	180 01.3	N10 01.5	220 46.7	15.4	S12 22.7	12.5	54.6
01	195 01.5	02.4	235 21.1	15.4	12 10.2	12.5	54.6
02	210 01.6	03.3	249 55.5	15.5	11 57.7	12.5	54.6
03	225 01.8	. . 04.2	264 30.0	15.6	11 45.2	12.6	54.6
04	240 01.9	05.1	279 04.6	15.6	11 32.6	12.7	54.5
05	255 02.1	06.0	293 39.2	15.7	11 19.9	12.6	54.5
06	270 02.2	N10 06.9	308 13.9	15.7	S11 07.3	12.7	54.5
07	285 02.4	07.7	322 48.6	15.8	10 54.6	12.8	54.5
08	300 02.5	08.6	337 23.4	15.8	10 41.8	12.8	54.5
09	315 02.7	. . 09.5	351 58.2	15.9	10 29.0	12.8	54.5
10	330 02.8	10.4	6 33.1	15.9	10 16.2	12.8	54.5
11	345 03.0	11.3	21 08.0	16.0	10 03.4	12.9	54.4
12	0 03.1	N10 12.2	35 43.0	16.0	S 9 50.5	12.9	54.4
13	15 03.3	13.1	50 18.0	16.1	9 37.6	13.0	54.4
14	30 03.4	13.9	64 53.1	16.1	9 24.6	12.9	54.4
15	45 03.6	. . 14.8	79 28.2	16.1	9 11.7	13.0	54.4
16	60 03.7	15.7	94 03.3	16.2	8 58.7	13.1	54.4
17	75 03.8	16.6	108 38.5	16.3	8 45.6	13.0	54.4
18	90 04.0	N10 17.5	123 13.8	16.3	S 8 32.6	13.1	54.4
19	105 04.1	18.4	137 49.1	16.3	8 19.5	13.1	54.3
20	120 04.3	19.2	152 24.4	16.3	8 06.4	13.2	54.3
21	135 04.4	. . 20.1	166 59.7	16.4	7 53.2	13.1	54.3
22	150 04.6	21.0	181 35.1	16.5	7 40.1	13.2	54.3
23	165 04.7	21.9	196 10.6	16.5	7 26.9	13.2	54.3
17 00	180 04.9	N10 22.8	210 46.1	16.5	S 7 13.7	13.2	54.3
01	195 05.0	23.7	225 21.6	16.5	7 00.5	13.3	54.3
02	210 05.2	24.5	239 57.1	16.6	6 47.2	13.3	54.3
03	225 05.3	. . 25.4	254 32.7	16.6	6 33.9	13.2	54.3
04	240 05.4	26.3	269 08.3	16.6	6 20.7	13.4	54.2
05	255 05.6	27.2	283 43.9	16.7	6 07.3	13.3	54.2
06	270 05.7	N10 28.1	298 19.6	16.7	S 5 54.0	13.3	54.2
07	285 05.9	28.9	312 55.3	16.7	5 40.7	13.4	54.2
08	300 06.0	29.8	327 31.0	16.8	5 27.3	13.4	54.2
09	315 06.2	. . 30.7	342 06.8	16.8	5 13.9	13.4	54.2
10	330 06.3	31.6	356 42.6	16.8	5 00.5	13.4	54.2
11	345 06.5	32.5	11 18.4	16.8	4 47.1	13.4	54.2
12	0 06.6	N10 33.3	25 54.2	16.9	S 4 33.7	13.4	54.2
13	15 06.7	34.2	40 30.1	16.8	4 20.3	13.5	54.2
14	30 06.9	35.1	55 05.9	16.9	4 06.8	13.5	54.1
15	45 07.0	. . 36.0	69 41.8	17.0	3 53.3	13.4	54.1
16	60 07.2	36.8	84 17.8	16.9	3 39.9	13.5	54.1
17	75 07.3	37.7	98 53.7	17.0	3 26.4	13.5	54.1
18	90 07.5	N10 38.6	113 29.7	16.9	S 3 12.9	13.5	54.1
19	105 07.6	39.5	128 05.6	17.0	2 59.4	13.5	54.1
20	120 07.7	40.4	142 41.6	17.0	2 45.9	13.5	54.1
21	135 07.9	. . 41.2	157 17.6	17.0	2 32.4	13.5	54.1
22	150 08.0	42.1	171 53.6	17.1	2 18.9	13.6	54.1
23	165 08.2	43.0	186 29.7	17.0	2 05.3	13.5	54.1
18 00	180 08.3	N10 43.9	201 05.7	17.1	S 1 51.8	13.5	54.1
01	195 08.4	44.7	215 41.8	17.1	1 38.3	13.6	54.1
02	210 08.6	45.6	230 17.9	17.0	1 24.7	13.5	54.1
03	225 08.7	. . 46.5	244 53.9	17.1	1 11.2	13.6	54.1
04	240 08.9	47.4	259 30.0	17.1	0 57.6	13.5	54.0
05	255 09.0	48.2	274 06.1	17.1	0 44.1	13.6	54.0
06	270 09.1	N10 49.1	288 42.2	17.1	S 0 30.5	13.5	54.0
07	285 09.3	50.0	303 18.3	17.1	0 17.0	13.6	54.0
08	300 09.4	50.8	317 54.4	17.2	S 0 03.4	13.5	54.0
09	315 09.6	. . 51.7	332 30.6	17.1	N 0 10.1	13.5	54.0
10	330 09.7	52.6	347 06.7	17.1	0 23.6	13.6	54.0
11	345 09.8	53.5	1 42.8	17.1	0 37.2	13.5	54.0
12	0 10.0	N10 54.3	16 18.9	17.1	N 0 50.7	13.5	54.0
13	15 10.1	55.2	30 55.0	17.2	1 04.2	13.6	54.0
14	30 10.3	56.1	45 31.2	17.1	1 17.8	13.5	54.0
15	45 10.4	. . 56.9	60 07.3	17.1	1 31.3	13.5	54.0
16	60 10.5	57.8	74 43.4	17.1	1 44.8	13.5	54.0
17	75 10.7	58.7	89 19.5	17.1	1 58.3	13.5	54.0
18	90 10.8	N10 59.5	103 55.6	17.1	N 2 11.8	13.5	54.0
19	105 10.9	11 00.4	118 31.7	17.1	2 25.3	13.5	54.0
20	120 11.1	01.3	133 07.8	17.1	2 38.8	13.4	54.0
21	135 11.2	. . 02.2	147 43.9	17.1	2 52.2	13.5	54.0
22	150 11.4	03.0	162 20.0	17.1	3 05.7	13.4	54.0
23	165 11.5	03.9	176 56.1	17.0	3 19.1	13.5	54.0
	S.D. 16.0	d 0.9	S.D.	14.8	14.8		14.7

Lat. °	Twilight Naut. h m	Twilight Civil h m	Sunrise h m	Moonrise 16 h m	17 h m	18 h m	19 h m
N 72	////	01 38	03 28	05 58	05 23	04 54	04 26
N 70	////	02 21	03 46	05 40	05 15	04 53	04 32
68	////	02 49	04 01	05 25	05 08	04 52	04 37
66	01 25	03 10	04 13	05 13	05 02	04 51	04 41
64	02 03	03 26	04 22	05 03	04 57	04 50	04 44
62	02 29	03 40	04 31	04 55	04 52	04 50	04 47
60	02 48	03 51	04 38	04 47	04 49	04 49	04 50
N 58	03 04	04 01	04 44	04 41	04 45	04 49	04 53
56	03 17	04 09	04 50	04 35	04 42	04 49	04 55
54	03 28	04 17	04 55	04 30	04 40	04 48	04 57
52	03 38	04 23	04 59	04 25	04 37	04 48	04 58
50	03 47	04 29	05 03	04 21	04 35	04 48	05 00
45	04 04	04 41	05 12	04 12	04 30	04 47	05 04
N 40	04 18	04 51	05 20	04 04	04 26	04 46	05 07
35	04 29	05 00	05 26	03 57	04 22	04 46	05 09
30	04 38	05 07	05 31	03 51	04 19	04 45	05 12
20	04 52	05 18	05 41	03 41	04 13	04 45	05 16
N 10	05 02	05 27	05 49	03 32	04 09	04 44	05 19
0	05 11	05 35	05 56	03 23	04 04	04 44	05 23
S 10	05 18	05 42	06 04	03 14	03 59	04 43	05 26
20	05 23	05 49	06 11	03 05	03 54	04 42	05 30
30	05 28	05 56	06 20	02 55	03 49	04 42	05 34
35	05 30	06 00	06 25	02 49	03 45	04 41	05 37
40	05 32	06 03	06 31	02 42	03 42	04 41	05 40
45	05 33	06 08	06 37	02 33	03 37	04 40	05 43
S 50	05 35	06 12	06 45	02 24	03 32	04 40	05 47
52	05 35	06 14	06 49	02 19	03 30	04 39	05 49
54	05 35	06 16	06 53	02 14	03 27	04 39	05 51
56	05 36	06 19	06 57	02 08	03 24	04 39	05 53
58	05 36	06 21	07 02	02 02	03 21	04 38	05 55
S 60	05 36	06 24	07 07	01 55	03 18	04 38	05 58

Lat. °	Sunset h m	Twilight Civil h m	Twilight Naut. h m	Moonset 16 h m	17 h m	18 h m	19 h m
N 72	20 35	22 31	////	13 34	15 34	17 28	19 24
N 70	20 16	21 44	////	13 50	15 40	17 25	19 12
68	20 01	21 14	////	14 03	15 44	17 23	19 02
66	19 49	20 53	22 43	14 13	15 48	17 21	18 54
64	19 39	20 35	22 01	14 21	15 51	17 19	18 48
62	19 30	20 22	21 34	14 29	15 54	17 18	18 42
60	19 23	20 10	21 14	14 35	15 56	17 17	18 37
N 58	19 16	20 00	20 58	14 40	15 59	17 15	18 33
56	19 11	19 52	20 44	14 45	16 00	17 14	18 29
54	19 06	19 44	20 33	14 49	16 02	17 14	18 25
52	19 01	19 37	20 23	14 53	16 04	17 13	18 22
50	18 57	19 31	20 14	14 57	16 05	17 12	18 19
45	18 48	19 19	19 56	15 05	16 08	17 11	18 13
N 40	18 40	19 09	19 42	15 11	16 10	17 09	18 08
35	18 34	19 00	19 31	15 16	16 13	17 08	18 04
30	18 28	18 53	19 22	15 21	16 14	17 07	18 00
20	18 19	18 41	19 08	15 29	16 18	17 05	17 53
N 10	18 11	18 32	18 57	15 36	16 20	17 04	17 47
0	18 03	18 24	18 48	15 43	16 23	17 02	17 42
S 10	17 55	18 17	18 41	15 50	16 26	17 01	17 36
20	17 47	18 10	18 36	15 56	16 28	16 59	17 31
30	17 39	18 03	18 31	16 04	16 31	16 58	17 24
35	17 33	17 59	18 29	16 09	16 33	16 57	17 20
40	17 28	17 55	18 27	16 14	16 35	16 56	17 16
45	17 21	17 51	18 25	16 20	16 37	16 54	17 11
S 50	17 13	17 46	18 24	16 27	16 40	16 53	17 05
52	17 10	17 44	18 23	16 30	16 41	16 52	17 03
54	17 06	17 42	18 23	16 34	16 43	16 51	17 00
56	17 01	17 39	18 22	16 37	16 44	16 50	16 57
58	16 56	17 37	18 22	16 42	16 46	16 49	16 53
S 60	16 51	17 35	18 22	16 47	16 48	16 48	16 49

Day	SUN Eqn. of Time 00 h m s	12 h m s	Mer. Pass. h m	MOON Mer. Pass. Upper h m	Lower h m	Age d	Phase
16	00 05	00 12	12 00	09 33	21 53	26	●
17	00 19	00 26	12 00	10 14	22 33	27	
18	00 33	00 40	11 59	10 53	23 13	28	

1985 APRIL 19, 20, 21 (FRI., SAT., SUN.)

G.M.T. (UT) d h	ARIES G.H.A.	VENUS −3.9 G.H.A.	VENUS Dec.	MARS +1.7 G.H.A.	MARS Dec.	JUPITER −1.8 G.H.A.	JUPITER Dec.	SATURN +0.4 G.H.A.	SATURN Dec.
	° ′	° ′	° ′	° ′	° ′	° ′	° ′	° ′	° ′
19 00	207 02.3	202 46.7	N 7 05.0	154 40.0	N19 25.1	250 45.0	S17 11.1	332 03.4	S17 07.9
01	222 04.7	217 49.5	04.3	169 40.7	25.5	265 47.2	11.0	347 06.0	07.8
02	237 07.2	232 52.2	03.5	184 41.3	26.0	280 49.3	10.9	2 08.6	07.8
03	252 09.7	247 55.0	· · 02.7	199 42.0	· · 26.4	295 51.4	· · 10.8	17 11.2	· · 07.8
04	267 12.1	262 57.8	01.9	214 42.7	26.9	310 53.6	10.7	32 13.8	07.7
05	282 14.6	278 00.6	01.2	229 43.3	27.3	325 55.7	10.6	47 16.5	07.7
06	297 17.1	293 03.3	N 7 00.4	244 44.0	N19 27.8	340 57.8	S17 10.5	62 19.1	S17 07.7
07	312 19.5	308 06.1	6 59.6	259 44.7	28.2	356 00.0	10.5	77 21.7	07.6
08	327 22.0	323 08.8	58.9	274 45.3	28.7	11 02.1	10.4	92 24.3	07.6
F 09	342 24.5	338 11.6	· · 58.1	289 46.0	· · 29.1	26 04.2	· · 10.3	107 27.0	· · 07.5
R 10	357 26.9	353 14.4	57.4	304 46.7	29.6	41 06.4	10.2	122 29.6	07.5
I 11	12 29.4	8 17.1	56.6	319 47.3	30.0	56 08.5	10.1	137 32.2	07.5
D 12	27 31.8	23 19.8	N 6 55.8	334 48.0	N19 30.5	71 10.6	S17 10.0	152 34.8	S17 07.4
A 13	42 34.3	38 22.6	55.1	349 48.7	30.9	86 12.7	09.9	167 37.4	07.4
Y 14	57 36.8	53 25.3	54.3	4 49.3	31.4	101 14.9	09.9	182 40.1	07.3
15	72 39.2	68 28.0	· · 53.6	19 50.0	· · 31.8	116 17.0	· · 09.8	197 42.7	· · 07.3
16	87 41.7	83 30.8	52.8	34 50.7	32.2	131 19.1	09.7	212 45.3	07.3
17	102 44.2	98 33.5	52.1	49 51.3	32.7	146 21.3	09.6	227 47.9	07.2
18	117 46.6	113 36.2	N 6 51.3	64 52.0	N19 33.1	161 23.4	S17 09.5	242 50.5	S17 07.2
19	132 49.1	128 38.9	50.6	79 52.7	33.6	176 25.5	09.4	257 53.2	07.2
20	147 51.6	143 41.6	49.9	94 53.3	34.0	191 27.7	09.3	272 55.8	07.1
21	162 54.0	158 44.3	· · 49.1	109 54.0	· · 34.5	206 29.8	· · 09.2	287 58.4	· · 07.1
22	177 56.5	173 47.0	48.4	124 54.7	34.9	221 31.9	09.2	303 01.0	07.0
23	192 59.0	188 49.7	47.6	139 55.3	35.3	236 34.1	09.1	318 03.6	07.0
20 00	208 01.4	203 52.4	N 6 46.9	154 56.0	N19 35.8	251 36.2	S17 09.0	333 06.3	S17 07.0
01	223 03.9	218 55.1	46.2	169 56.7	36.2	266 38.3	08.9	348 08.9	06.9
02	238 06.3	233 57.8	45.4	184 57.3	36.7	281 40.5	08.8	3 11.5	06.9
03	253 08.8	249 00.5	· · 44.7	199 58.0	· · 37.1	296 42.6	· · 08.7	18 14.1	· · 06.8
04	268 11.3	264 03.2	44.0	214 58.7	37.5	311 44.8	08.6	33 16.8	06.8
05	283 13.7	279 05.9	43.2	229 59.3	38.0	326 46.9	08.6	48 19.4	06.8
06	298 16.2	294 08.5	N 6 42.5	245 00.0	N19 38.4	341 49.0	S17 08.5	63 22.0	S17 06.7
07	313 18.7	309 11.2	41.8	260 00.7	38.9	356 51.2	08.4	78 24.6	06.7
S 08	328 21.1	324 13.9	41.1	275 01.3	39.3	11 53.3	08.3	93 27.2	06.6
A 09	343 23.6	339 16.5	· · 40.3	290 02.0	· · 39.7	26 55.4	· · 08.2	108 29.9	· · 06.6
T 10	358 26.1	354 19.2	39.6	305 02.6	40.2	41 57.6	08.1	123 32.5	06.6
U 11	13 28.5	9 21.8	38.9	320 03.3	40.6	56 59.7	08.0	138 35.1	06.5
R 12	28 31.0	24 24.5	N 6 38.2	335 04.0	N19 41.0	72 01.9	S17 07.9	153 37.7	S17 06.5
D 13	43 33.5	39 27.1	37.5	350 04.6	41.5	87 04.0	07.9	168 40.4	06.5
A 14	58 35.9	54 29.8	36.7	5 05.3	41.9	102 06.1	07.8	183 43.0	06.4
Y 15	73 38.4	69 32.4	· · 36.0	20 06.0	· · 42.4	117 08.3	· · 07.7	198 45.6	· · 06.4
16	88 40.8	84 35.0	35.3	35 06.6	42.8	132 10.4	07.6	213 48.2	06.3
17	103 43.3	99 37.7	34.6	50 07.3	43.2	147 12.5	07.5	228 50.9	06.3
18	118 45.8	114 40.3	N 6 33.9	65 08.0	N19 43.7	162 14.7	S17 07.4	243 53.5	S17 06.3
19	133 48.2	129 42.9	33.2	80 08.6	44.1	177 16.8	07.3	258 56.1	06.2
20	148 50.7	144 45.5	32.5	95 09.3	44.5	192 19.0	07.3	273 58.7	06.2
21	163 53.2	159 48.1	· · 31.8	110 09.9	· · 45.0	207 21.1	· · 07.2	289 01.4	· · 06.1
22	178 55.6	174 50.8	31.1	125 10.6	45.4	222 23.2	07.1	304 04.0	06.1
23	193 58.1	189 53.4	30.4	140 11.3	45.8	237 25.4	07.0	319 06.6	06.1
21 00	209 00.6	204 56.0	N 6 29.7	155 11.9	N19 46.3	252 27.5	S17 06.9	334 09.2	S17 06.0
01	224 03.0	219 58.6	29.0	170 12.6	46.7	267 29.7	06.8	349 11.9	06.0
02	239 05.5	235 01.2	28.3	185 13.3	47.1	282 31.8	06.8	4 14.5	05.9
03	254 08.0	250 03.8	· · 27.6	200 13.9	· · 47.6	297 33.9	· · 06.7	19 17.1	· · 05.9
04	269 10.4	265 06.3	26.9	215 14.6	48.0	312 36.1	06.6	34 19.7	05.9
05	284 12.9	280 08.9	26.2	230 15.2	48.4	327 38.2	06.5	49 22.3	05.8
06	299 15.3	295 11.5	N 6 25.6	245 15.9	N19 48.9	342 40.4	S17 06.4	64 25.0	S17 05.8
07	314 17.8	310 14.1	24.9	260 16.6	49.3	357 42.5	06.3	79 27.6	05.7
08	329 20.3	325 16.6	24.2	275 17.2	49.7	12 44.7	06.2	94 30.2	05.7
S 09	344 22.7	340 19.2	· · 23.5	290 17.9	· · 50.2	27 46.8	· · 06.2	109 32.8	· · 05.7
U 10	359 25.2	355 21.8	22.8	305 18.6	50.6	42 48.9	06.1	124 35.5	05.6
N 11	14 27.7	10 24.3	22.2	320 19.2	51.0	57 51.1	06.0	139 38.1	05.6
D 12	29 30.1	25 26.9	N 6 21.5	335 19.9	N19 51.5	72 53.2	S17 05.9	154 40.7	S17 05.6
A 13	44 32.6	40 29.4	20.8	350 20.5	51.9	87 55.4	05.8	169 43.4	05.5
Y 14	59 35.1	55 32.0	20.1	5 21.2	52.3	102 57.5	05.7	184 46.0	05.5
15	74 37.5	70 34.5	· · 19.5	20 21.9	· · 52.7	117 59.7	· · 05.7	199 48.6	· · 05.4
16	89 40.0	85 37.1	18.8	35 22.5	53.2	133 01.8	05.6	214 51.2	05.4
17	104 42.4	100 39.6	18.1	50 23.2	53.6	148 04.0	05.5	229 53.9	05.4
18	119 44.9	115 42.1	N 6 17.5	65 23.9	N19 54.0	163 06.1	S17 05.4	244 56.5	S17 05.3
19	134 47.4	130 44.7	16.8	80 24.5	54.5	178 08.2	05.3	259 59.1	05.3
20	149 49.8	145 47.2	16.1	95 25.2	54.9	193 10.4	05.2	275 01.7	05.2
21	164 52.3	160 49.7	· · 15.5	110 25.8	· · 55.3	208 12.5	· · 05.2	290 04.4	· · 05.2
22	179 54.8	175 52.2	14.8	125 26.5	55.7	223 14.7	05.1	305 07.0	05.2
23	194 57.2	190 54.8	14.2	140 27.2	56.2	238 16.8	05.0	320 09.6	05.1
Mer. Pass.	h m 10 06.2	v 2.6	d 0.7	v 0.7	d 0.4	v 2.1	d 0.1	v 2.6	d 0.0

STARS

Name	S.H.A.	Dec.
	° ′	° ′
Acamar	315 35.0	S40 21.9
Achernar	335 43.1	S57 18.7
Acrux	173 33.2	S63 01.2
Adhara	255 29.6	S28 57.2
Aldebaran	291 14.4	N16 28.8
Alioth	166 38.7	N56 02.5
Alkaid	153 15.1	N49 23.2
Al Na'ir	28 10.7	S47 02.0
Alnilam	276 08.4	S 1 12.7
Alphard	218 17.1	S 8 35.7
Alphecca	126 28.9	N26 45.6
Alpheratz	358 06.2	N29 00.3
Altair	62 29.2	N 8 49.4
Ankaa	353 37.1	S42 23.2
Antares	112 52.4	S26 24.1
Arcturus	146 15.0	N19 15.4
Atria	108 13.3	S69 00.1
Avior	234 27.0	S59 27.9
Bellatrix	278 55.3	N 6 20.2
Betelgeuse	271 24.8	N 7 24.3
Canopus	264 06.0	S52 41.4
Capella	281 06.6	N45 59.2
Deneb	49 46.3	N45 13.2
Denebola	182 55.3	N14 39.3
Diphda	349 17.7	S18 04.2
Dubhe	194 17.2	N61 50.1
Elnath	278 40.1	N28 35.9
Eltanin	90 55.9	N51 29.0
Enif	34 08.4	N 9 48.1
Fomalhaut	15 47.8	S29 42.1
Gacrux	172 24.7	S57 02.0
Gienah	176 14.2	S17 27.7
Hadar	149 18.2	S60 18.2
Hamal	328 25.5	N23 23.5
Kaus Aust.	84 12.2	S34 23.6
Kochab	137 17.5	N74 12.8
Markab	14 00.1	N15 07.3
Menkar	314 37.9	N 4 01.9
Menkent	148 32.7	S36 18.0
Miaplacidus	221 44.3	S69 39.6
Mirfak	309 11.8	N49 48.6
Nunki	76 24.8	S26 19.0
Peacock	53 52.9	S56 46.9
Pollux	243 54.0	N28 03.9
Procyon	245 22.3	N 5 15.9
Rasalhague	96 26.2	N12 33.9
Regulus	208 06.2	N12 02.4
Rigel	281 33.0	S 8 13.1
Rigil Kent.	140 20.8	S60 46.5
Sabik	102 37.0	S15 42.6
Schedar	350 05.9	N56 27.2
Shaula	96 50.9	S37 05.7
Sirius	258 52.8	S16 41.8
Spica	158 53.7	S11 05.2
Suhail	223 08.3	S43 22.5
Vega	80 53.4	N38 45.8
Zuben'ubi	137 29.0	S15 59.0

	S.H.A.	Mer. Pass.
	° ′	h m
Venus	355 51.0	10 23
Mars	306 54.6	13 40
Jupiter	43 34.8	7 13
Saturn	125 04.9	1 47

1985 APRIL 19, 20, 21 (FRI., SAT., SUN.)

G.M.T. (UT) d h		SUN G.H.A. ° ′	SUN Dec. ° ′	MOON G.H.A. ° ′	v ′	MOON Dec. ° ′	d ′	H.P. ′
19 00	FRIDAY	180 11.6	N11 04.8	191 32.1	17.1	N 3 32.6	13.4	54.0
01		195 11.8	05.6	206 08.2	17.0	3 46.0	13.4	54.0
02		210 11.9	06.5	220 44.2	17.1	3 59.4	13.4	54.0
03		225 12.0	.. 07.4	235 20.3	17.0	4 12.8	13.4	54.0
04		240 12.2	08.2	249 56.3	17.0	4 26.2	13.3	54.0
05		255 12.3	09.1	264 32.3	17.0	4 39.5	13.4	53.9
06		270 12.5	N11 10.0	279 08.3	16.9	N 4 52.9	13.3	53.9
07		285 12.6	10.8	293 44.2	17.0	5 06.2	13.3	53.9
08		300 12.7	11.7	308 20.2	16.9	5 19.5	13.3	53.9
09		315 12.9	.. 12.5	322 56.1	16.9	5 32.8	13.3	53.9
10		330 13.0	13.4	337 32.0	16.9	5 46.1	13.2	53.9
11		345 13.1	14.3	352 07.9	16.9	5 59.3	13.2	53.9
12		0 13.3	N11 15.1	6 43.8	16.8	N 6 12.5	13.3	53.9
13		15 13.4	16.0	21 19.6	16.8	6 25.8	13.1	53.9
14		30 13.5	16.9	35 55.4	16.8	6 38.9	13.2	53.9
15		45 13.7	.. 17.7	50 31.2	16.8	6 52.1	13.1	53.9
16		60 13.8	18.6	65 07.0	16.8	7 05.2	13.1	53.9
17		75 13.9	19.5	79 42.8	16.7	7 18.3	13.1	53.9
18		90 14.1	N11 20.3	94 18.5	16.7	N 7 31.4	13.1	53.9
19		105 14.2	21.2	108 54.2	16.7	7 44.5	13.0	53.9
20		120 14.3	22.0	123 29.9	16.6	7 57.5	13.0	53.9
21		135 14.5	.. 22.9	138 05.5	16.6	8 10.5	13.0	53.9
22		150 14.6	23.8	152 41.1	16.6	8 23.5	12.9	53.9
23		165 14.7	24.6	167 16.7	16.5	8 36.4	12.9	53.9
20 00	SATURDAY	180 14.9	N11 25.5	181 52.2	16.5	N 8 49.3	12.9	53.9
01		195 15.0	26.3	196 27.7	16.5	9 02.2	12.9	53.9
02		210 15.1	27.2	211 03.2	16.5	9 15.1	12.8	53.9
03		225 15.3	.. 28.1	225 38.7	16.4	9 27.9	12.8	53.9
04		240 15.4	28.9	240 14.1	16.4	9 40.7	12.7	53.9
05		255 15.5	29.8	254 49.5	16.3	9 53.4	12.7	53.9
06		270 15.7	N11 30.6	269 24.8	16.3	N10 06.1	12.7	53.9
07		285 15.8	31.5	284 00.1	16.3	10 18.8	12.6	54.0
08		300 15.9	32.3	298 35.4	16.2	10 31.4	12.6	54.0
09		315 16.1	.. 33.2	313 10.6	16.2	10 44.0	12.6	54.0
10		330 16.2	34.1	327 45.8	16.1	10 56.6	12.5	54.0
11		345 16.3	34.9	342 20.9	16.1	11 09.1	12.5	54.0
12		0 16.4	N11 35.8	356 56.0	16.1	N11 21.6	12.4	54.0
13		15 16.6	36.6	11 31.1	16.0	11 34.0	12.5	54.0
14		30 16.7	37.5	26 06.1	16.0	11 46.5	12.3	54.0
15		45 16.8	.. 38.3	40 41.1	15.9	11 58.8	12.3	54.0
16		60 17.0	39.2	55 16.0	15.9	12 11.1	12.3	54.0
17		75 17.1	40.0	69 50.9	15.8	12 23.4	12.2	54.0
18		90 17.2	N11 40.9	84 25.7	15.8	N12 35.6	12.2	54.0
19		105 17.4	41.7	99 00.5	15.7	12 47.8	12.1	54.0
20		120 17.5	42.6	113 35.2	15.7	12 59.9	12.1	54.0
21		135 17.6	.. 43.5	128 09.9	15.7	13 12.0	12.1	54.0
22		150 17.7	44.3	142 44.6	15.6	13 24.1	12.0	54.0
23		165 17.9	45.2	157 19.2	15.5	13 36.1	11.9	54.0
21 00	SUNDAY	180 18.0	N11 46.0	171 53.7	15.5	N13 48.0	11.9	54.0
01		195 18.1	46.9	186 28.2	15.4	13 59.9	11.8	54.0
02		210 18.3	47.7	201 02.6	15.4	14 11.7	11.8	54.0
03		225 18.4	.. 48.6	215 37.0	15.3	14 23.5	11.8	54.0
04		240 18.5	49.4	230 11.3	15.3	14 35.3	11.6	54.0
05		255 18.6	50.3	244 45.6	15.2	14 46.9	11.7	54.0
06		270 18.8	N11 51.1	259 19.8	15.2	N14 58.6	11.5	54.0
07		285 18.9	52.0	273 54.0	15.1	15 10.1	11.5	54.0
08		300 19.0	52.8	288 28.1	15.0	15 21.6	11.5	54.1
09		315 19.1	.. 53.7	303 02.1	15.0	15 33.1	11.4	54.1
10		330 19.3	54.5	317 36.1	15.0	15 44.5	11.3	54.1
11		345 19.4	55.4	332 10.1	14.8	15 55.8	11.3	54.1
12		0 19.5	N11 56.2	346 43.9	14.9	N16 07.1	11.2	54.1
13		15 19.6	57.0	1 17.8	14.7	16 18.3	11.2	54.1
14		30 19.8	57.9	15 51.5	14.7	16 29.5	11.0	54.1
15		45 19.9	.. 58.7	30 25.2	14.6	16 40.5	11.1	54.1
16		60 20.0	11 59.6	44 58.8	14.6	16 51.6	10.9	54.1
17		75 20.1	12 00.4	59 32.4	14.5	17 02.5	10.9	54.1
18		90 20.3	N12 01.3	74 05.9	14.5	N17 13.4	10.8	54.1
19		105 20.4	02.1	88 39.4	14.3	17 24.2	10.8	54.1
20		120 20.5	03.0	103 12.7	14.4	17 35.0	10.7	54.1
21		135 20.6	.. 03.8	117 46.1	14.2	17 45.7	10.6	54.1
22		150 20.8	04.7	132 19.3	14.2	17 56.3	10.6	54.1
23		165 20.9	05.5	146 52.5	14.1	18 06.9	10.5	54.2
		S.D. 15.9	d 0.9	S.D. 14.7		14.7		14.7

Lat. °	Twilight Naut. h m	Twilight Civil h m	Sunrise h m	Moonrise 19 h m	Moonrise 20 h m	Moonrise 21 h m	Moonrise 22 h m
N 72	////	00 57	03 10	04 26	03 54	03 10	▭
N 70	////	01 58	03 31	04 32	04 08	03 39	02 46
68	////	02 32	03 48	04 37	04 20	04 00	03 31
66	00 49	02 55	04 01	04 41	04 30	04 17	04 01
64	01 43	03 14	04 12	04 44	04 38	04 31	04 23
62	02 13	03 29	04 21	04 47	04 45	04 43	04 41
60	02 36	03 41	04 29	04 50	04 51	04 53	04 56
N 58	02 53	03 52	04 36	04 53	04 57	05 02	05 09
56	03 08	04 01	04 42	04 55	05 01	05 10	05 20
54	03 20	04 09	04 48	04 57	05 06	05 17	05 30
52	03 30	04 16	04 53	04 58	05 10	05 23	05 39
50	03 39	04 23	04 57	05 00	05 13	05 29	05 47
45	03 58	04 36	05 07	05 04	05 21	05 41	06 04
N 40	04 13	04 47	05 15	05 07	05 28	05 51	06 18
35	04 24	04 56	05 22	05 09	05 34	06 00	06 29
30	04 34	05 03	05 28	05 12	05 39	06 08	06 40
20	04 49	05 16	05 38	05 16	05 48	06 21	06 58
N 10	05 01	05 26	05 47	05 19	05 55	06 33	07 13
0	05 10	05 34	05 56	05 23	06 03	06 44	07 28
S 10	05 18	05 42	06 04	05 26	06 10	06 56	07 43
20	05 24	05 50	06 12	05 30	06 18	07 08	07 59
30	05 30	05 58	06 22	05 34	06 27	07 22	08 17
35	05 32	06 02	06 28	05 37	06 33	07 30	08 28
40	05 35	06 06	06 34	05 40	06 39	07 39	08 41
45	05 37	06 11	06 41	05 43	06 46	07 50	08 55
S 50	05 39	06 17	06 50	05 47	06 54	08 03	09 14
52	05 40	06 19	06 54	05 49	06 58	08 10	09 22
54	05 40	06 22	06 58	05 51	07 03	08 16	09 32
56	05 41	06 25	07 03	05 53	07 08	08 24	09 43
58	05 42	06 28	07 08	05 55	07 13	08 33	09 55
S 60	05 43	06 31	07 14	05 58	07 19	08 43	10 10

Lat. °	Sunset h m	Twilight Civil h m	Twilight Naut. h m	Moonset 19 h m	Moonset 20 h m	Moonset 21 h m	Moonset 22 h m
N 72	20 52	23 21	////	19 24	21 34	▭	▭
N 70	20 30	22 07	////	19 12	21 07	23 29	▭
68	20 13	21 31	////	19 02	20 47	22 46	▭
66	19 59	21 06	23 26	18 54	20 32	22 17	24 21
64	19 48	20 47	22 21	18 48	20 19	21 56	23 41
62	19 39	20 31	21 49	18 42	20 08	21 38	23 13
60	19 30	20 19	21 25	18 37	19 59	21 24	22 52
N 58	19 23	20 08	21 07	18 33	19 51	21 12	22 35
56	19 17	19 58	20 53	18 29	19 44	21 02	22 21
54	19 11	19 50	20 40	18 25	19 38	20 52	22 08
52	19 06	19 43	20 29	18 22	19 32	20 44	21 57
50	19 01	19 36	20 20	18 19	19 27	20 37	21 47
45	18 52	19 23	20 01	18 13	19 17	20 21	21 27
N 40	18 43	19 12	19 46	18 08	19 08	20 09	21 11
35	18 36	19 03	19 34	18 04	19 00	19 58	20 57
30	18 30	18 55	19 24	18 00	18 53	19 48	20 45
20	18 20	18 42	19 09	17 53	18 42	19 32	20 25
N 10	18 11	18 32	18 57	17 47	18 32	19 18	20 07
0	18 02	18 23	18 48	17 42	18 23	19 05	19 50
S 10	17 54	18 15	18 40	17 36	18 13	18 52	19 34
20	17 45	18 08	18 34	17 31	18 03	18 38	19 16
30	17 35	18 00	18 28	17 24	17 52	18 22	18 56
35	17 30	17 56	18 25	17 20	17 46	18 13	18 45
40	17 23	17 51	18 23	17 16	17 38	18 03	18 31
45	17 16	17 46	18 20	17 11	17 30	17 51	18 16
S 50	17 07	17 40	18 18	17 05	17 20	17 36	17 56
52	17 03	17 38	18 17	17 03	17 15	17 29	17 47
54	16 59	17 35	18 16	17 00	17 10	17 22	17 37
56	16 54	17 32	18 16	16 57	17 04	17 13	17 26
58	16 48	17 29	18 15	16 53	16 58	17 04	17 12
S 60	16 42	17 26	18 14	16 49	16 50	16 53	16 57

Day	SUN Eqn. of Time 00h m s	SUN Eqn. of Time 12h m s	SUN Mer. Pass. h m	MOON Mer. Pass. Upper h m	MOON Mer. Pass. Lower h m	Age d	Phase
19	00 46	00 53	11 59	11 32	23 52	29	●
20	00 59	01 06	11 59	12 13	24 33	00	
21	01 12	01 18	11 59	12 55	00 33	01	

1985 APRIL 22, 23, 24 (MON., TUES., WED.)

	G.M.T. (UT) d h	ARIES G.H.A.	VENUS −4.0 G.H.A.	VENUS Dec.	MARS +1.7 G.H.A.	MARS Dec.	JUPITER −1.8 G.H.A.	JUPITER Dec.	SATURN +0.4 G.H.A.	SATURN Dec.
		° ′	° ′	° ′	° ′	° ′	° ′	° ′	° ′	° ′
	22 00	209 59.7	205 57.3	N 6 13.5	155 27.8	N19 56.6	253 19.0	S17 04.9	335 12.2	S17 05.1
	01	225 02.2	220 59.8	12.9	170 28.5	57.0	268 21.1	04.8	350 14.9	05.0
	02	240 04.6	236 02.3	12.2	185 29.1	57.4	283 23.3	04.7	5 17.5	05.0
	03	255 07.1	251 04.8	· · 11.6	200 29.8	· · 57.9	298 25.4	· · 04.7	20 20.1	· · 05.0
	04	270 09.6	266 07.3	10.9	215 30.5	58.3	313 27.6	04.6	35 22.7	04.9
	05	285 12.0	281 09.8	10.3	230 31.1	58.7	328 29.7	04.5	50 25.4	04.9
	06	300 14.5	296 12.2	N 6 09.6	245 31.8	N19 59.1	343 31.9	S17 04.4	65 28.0	S17 04.8
	07	315 16.9	311 14.7	09.0	260 32.4	19 59.6	358 34.0	04.3	80 30.6	04.8
	08	330 19.4	326 17.2	08.3	275 33.1	20 00.0	13 36.2	04.2	95 33.3	04.8
M	09	345 21.9	341 19.7	· · 07.7	290 33.8	· · 00.4	28 38.3	· · 04.2	110 35.9	· · 04.7
O	10	0 24.3	356 22.2	07.1	305 34.4	00.8	43 40.5	04.1	125 38.5	04.7
N	11	15 26.8	11 24.6	06.4	320 35.1	01.3	58 42.6	04.0	140 41.1	04.6
D	12	30 29.3	26 27.1	N 6 05.8	335 35.7	N20 01.7	73 44.8	S17 03.9	155 43.8	S17 04.6
A	13	45 31.7	41 29.5	05.2	350 36.4	02.1	88 46.9	03.8	170 46.4	04.6
Y	14	60 34.2	56 32.0	04.5	5 37.1	02.5	103 49.1	03.7	185 49.0	04.5
	15	75 36.7	71 34.5	· · 03.9	20 37.7	· · 02.9	118 51.2	· · 03.7	200 51.6	· · 04.5
	16	90 39.1	86 36.9	03.3	35 38.4	03.4	133 53.4	03.6	215 54.3	04.4
	17	105 41.6	101 39.3	02.7	50 39.0	03.8	148 55.5	03.5	230 56.9	04.4
	18	120 44.1	116 41.8	N 6 02.0	65 39.7	N20 04.2	163 57.7	S17 03.4	245 59.5	S17 04.4
	19	135 46.5	131 44.2	01.4	80 40.4	04.6	178 59.8	03.3	261 02.2	04.3
	20	150 49.0	146 46.7	00.8	95 41.0	05.0	194 02.0	03.2	276 04.8	04.3
	21	165 51.4	161 49.1	6 00.2	110 41.7	· · 05.5	209 04.1	· · 03.2	291 07.4	· · 04.2
	22	180 53.9	176 51.5	5 59.6	125 42.3	05.9	224 06.3	03.1	306 10.0	04.2
	23	195 56.4	191 53.9	59.0	140 43.0	06.3	239 08.4	03.0	321 12.7	04.2
	23 00	210 58.8	206 56.4	N 5 58.3	155 43.7	N20 06.7	254 10.6	S17 02.9	336 15.3	S17 04.1
	01	226 01.3	221 58.8	57.7	170 44.3	07.1	269 12.7	02.8	351 17.9	04.1
	02	241 03.8	237 01.2	57.1	185 45.0	07.6	284 14.9	02.8	6 20.6	04.0
	03	256 06.2	252 03.6	· · 56.5	200 45.6	· · 08.0	299 17.0	· · 02.7	21 23.2	· · 04.0
	04	271 08.7	267 06.0	55.9	215 46.3	08.4	314 19.2	02.6	36 25.8	04.0
	05	286 11.2	282 08.4	55.3	230 47.0	08.8	329 21.3	02.5	51 28.5	03.9
	06	301 13.6	297 10.8	N 5 54.7	245 47.6	N20 09.2	344 23.5	S17 02.4	66 31.1	S17 03.9
	07	316 16.1	312 13.2	54.1	260 48.3	09.6	359 25.6	02.3	81 33.7	03.8
T	08	331 18.5	327 15.6	53.5	275 48.9	10.0	14 27.8	02.3	96 36.3	03.8
U	09	346 21.0	342 18.0	· · 52.9	290 49.6	· · 10.5	29 30.0	· · 02.2	111 39.0	· · 03.8
E	10	1 23.5	357 20.3	52.3	305 50.2	10.9	44 32.1	02.1	126 41.6	03.7
S	11	16 25.9	12 22.7	51.7	320 50.9	11.3	59 34.3	02.0	141 44.2	03.7
D	12	31 28.4	27 25.1	N 5 51.2	335 51.6	N20 11.7	74 36.4	S17 01.9	156 46.9	S17 03.6
A	13	46 30.9	42 27.5	50.6	350 52.2	12.1	89 38.6	01.9	171 49.5	03.6
Y	14	61 33.3	57 29.8	50.0	5 52.9	12.5	104 40.7	01.8	186 52.1	03.6
	15	76 35.8	72 32.2	· · 49.4	20 53.5	· · 12.9	119 42.9	· · 01.7	201 54.7	· · 03.5
	16	91 38.3	87 34.5	48.8	35 54.2	13.4	134 45.0	01.6	216 57.4	03.5
	17	106 40.7	102 36.9	48.2	50 54.9	13.8	149 47.2	01.5	232 00.0	03.4
	18	121 43.2	117 39.2	N 5 47.7	65 55.5	N20 14.2	164 49.4	S17 01.5	247 02.6	S17 03.4
	19	136 45.7	132 41.6	47.1	80 56.2	14.6	179 51.5	01.4	262 05.3	03.4
	20	151 48.1	147 43.9	46.5	95 56.8	15.0	194 53.7	01.3	277 07.9	03.3
	21	166 50.6	162 46.3	· · 45.9	110 57.5	· · 15.4	209 55.8	· · 01.2	292 10.5	· · 03.3
	22	181 53.0	177 48.6	45.4	125 58.1	15.8	224 58.0	01.1	307 13.2	03.2
	23	196 55.5	192 50.9	44.8	140 58.8	16.2	240 00.1	01.1	322 15.8	03.2
	24 00	211 58.0	207 53.3	N 5 44.2	155 59.5	N20 16.7	255 02.3	S17 01.0	337 18.4	S17 03.2
	01	227 00.4	222 55.6	43.7	171 00.1	17.1	270 04.5	00.9	352 21.1	03.1
	02	242 02.9	237 57.9	43.1	186 00.8	17.5	285 06.6	00.8	7 23.7	03.1
	03	257 05.4	253 00.2	· · 42.5	201 01.4	· · 17.9	300 08.8	· · 00.7	22 26.3	· · 03.0
	04	272 07.8	268 02.5	42.0	216 02.1	18.3	315 10.9	00.7	37 28.9	03.0
	05	287 10.3	283 04.9	41.4	231 02.7	18.7	330 13.1	00.6	52 31.6	03.0
	06	302 12.8	298 07.2	N 5 40.9	246 03.4	N20 19.1	345 15.3	S17 00.5	67 34.2	S17 02.9
W	07	317 15.2	313 09.5	40.3	261 04.1	19.5	0 17.4	00.4	82 36.8	02.9
E	08	332 17.7	328 11.8	39.8	276 04.7	19.9	15 19.6	00.3	97 39.5	02.8
D	09	347 20.2	343 14.1	· · 39.2	291 05.4	· · 20.3	30 21.7	· · 00.3	112 42.1	· · 02.8
N	10	2 22.6	358 16.4	38.7	306 06.0	20.7	45 23.9	00.2	127 44.7	02.8
E	11	17 25.1	13 18.6	38.1	321 06.7	21.2	60 26.1	00.1	142 47.4	02.7
S	12	32 27.5	28 20.9	N 5 37.6	336 07.3	N20 21.6	75 28.2	S17 00.0	157 50.0	S17 02.7
D	13	47 30.0	43 23.2	37.0	351 08.0	22.0	90 30.4	16 59.9	172 52.6	02.6
A	14	62 32.5	58 25.5	36.5	6 08.6	22.4	105 32.6	59.9	187 55.3	02.6
Y	15	77 34.9	73 27.8	· · 36.0	21 09.3	· · 22.8	120 34.7	· · 59.8	202 57.9	· · 02.6
	16	92 37.4	88 30.0	35.4	36 10.0	23.2	135 36.9	59.7	218 00.5	02.5
	17	107 39.9	103 32.3	34.9	51 10.6	23.6	150 39.0	59.6	233 03.2	02.5
	18	122 42.3	118 34.6	N 5 34.3	66 11.3	N20 24.0	165 41.2	S16 59.5	248 05.8	S17 02.4
	19	137 44.8	133 36.8	33.8	81 11.9	24.4	180 43.4	59.5	263 08.4	02.4
	20	152 47.3	148 39.1	33.3	96 12.6	24.8	195 45.5	59.4	278 11.1	02.4
	21	167 49.7	163 41.3	· · 32.8	111 13.2	· · 25.2	210 47.7	· · 59.3	293 13.7	· · 02.3
	22	182 52.2	178 43.6	32.2	126 13.9	25.6	225 49.9	59.2	308 16.3	02.3
	23	197 54.6	193 45.8	31.7	141 14.5	26.0	240 52.0	59.1	323 19.0	02.2
	Mer. Pass.	h m 9 54.5	v 2.4	d 0.6	v 0.7	d 0.4	v 2.2	d 0.1	v 2.6	d 0.0

STARS

Name	S.H.A.	Dec.
	° ′	° ′
Acamar	315 35.0	S40 21.9
Achernar	335 43.1	S57 18.7
Acrux	173 33.2	S63 01.2
Adhara	255 29.6	S28 57.2
Aldebaran	291 14.4	N16 28.8
Alioth	166 38.7	N56 02.5
Alkaid	153 15.1	N49 23.2
Al Na'ir	28 10.7	S47 02.0
Alnilam	276 08.4	S 1 12.7
Alphard	218 17.2	S 8 35.7
Alphecca	126 28.8	N26 45.6
Alpheratz	358 06.2	N29 00.3
Altair	62 29.2	N 8 49.4
Ankaa	353 37.1	S42 23.2
Antares	112 52.4	S26 24.1
Arcturus	146 15.0	N19 15.4
Atria	108 13.3	S69 00.1
Avior	234 27.0	S59 27.9
Bellatrix	278 55.3	N 6 20.2
Betelgeuse	271 24.8	N 7 24.3
Canopus	264 06.0	S52 41.4
Capella	281 06.6	N45 59.2
Deneb	49 46.3	N45 13.2
Denebola	182 55.3	N14 39.3
Diphda	349 17.7	S18 04.2
Dubhe	194 17.2	N61 50.1
Elnath	278 40.1	N28 35.9
Eltanin	90 55.8	N51 29.0
Enif	34 08.4	N 9 48.1
Fomalhaut	15 47.8	S29 42.1
Gacrux	172 24.7	S57 02.0
Gienah	176 14.2	S17 27.7
Hadar	149 18.1	S60 18.2
Hamal	328 25.5	N23 23.5
Kaus Aust.	84 12.1	S34 23.6
Kochab	137 17.5	N74 12.9
Markab	14 00.0	N15 07.3
Menkar	314 37.9	N 4 01.9
Menkent	148 32.7	S36 18.0
Miaplacidus	221 44.4	S69 39.6
Mirfak	309 11.8	N49 48.6
Nunki	76 24.8	S26 19.0
Peacock	53 52.9	S56 46.9
Pollux	243 54.0	N28 03.9
Procyon	245 22.3	N 5 15.9
Rasalhague	96 26.2	N12 33.9
Regulus	208 06.2	N12 02.4
Rigel	281 33.0	S 8 13.1
Rigil Kent.	140 20.8	S60 46.5
Sabik	102 37.0	S15 42.6
Schedar	350 05.9	N56 27.2
Shaula	96 50.9	S37 05.7
Sirius	258 52.8	S16 41.8
Spica	158 53.7	S11 05.2
Suhail	223 08.3	S43 22.5
Vega	80 53.4	N38 45.8
Zuben'ubi	137 29.0	S15 59.0

	S.H.A.	Mer. Pass.
	° ′	h m
Venus	355 57.5	10 11
Mars	304 44.8	13 36
Jupiter	43 11.7	7 02
Saturn	125 16.5	1 35

1985 APRIL 22, 23, 24 (MON., TUES., WED.)

G.M.T. (UT) d	h	SUN G.H.A. ° ′	SUN Dec. ° ′	MOON G.H.A. ° ′	MOON *v* ′	MOON Dec. ° ′	MOON *d* ′	MOON H.P. ′
22	00	180 21.0	N12 06.3	161 25.6	14.1	N18 17.4	10.4	54.2
	01	195 21.1	07.2	175 58.7	14.0	18 27.8	10.3	54.2
	02	210 21.3	08.0	190 31.7	13.9	18 38.1	10.3	54.2
	03	225 21.4	. . 08.9	205 04.6	13.8	18 48.4	10.2	54.2
	04	240 21.5	09.7	219 37.4	13.8	18 58.6	10.1	54.2
	05	255 21.6	10.6	234 10.2	13.7	19 08.7	10.0	54.2
	06	270 21.8	N12 11.4	248 42.9	13.7	N19 18.7	10.0	54.2
	07	285 21.9	12.2	263 15.6	13.5	19 28.7	09.8	54.2
	08	300 22.0	13.1	277 48.1	13.5	19 38.5	09.8	54.2
M	09	315 22.1	. . 13.9	292 20.6	13.5	19 48.3	09.8	54.2
O	10	330 22.2	14.8	306 53.1	13.3	19 58.1	09.6	54.3
N	11	345 22.4	15.6	321 25.4	13.3	20 07.7	09.5	54.3
D	12	0 22.5	N12 16.4	335 57.7	13.3	N20 17.2	09.5	54.3
A	13	15 22.6	17.3	350 30.0	13.1	20 26.7	09.4	54.3
Y	14	30 22.7	18.1	5 02.1	13.1	20 36.1	09.3	54.3
	15	45 22.9	. . 19.0	19 34.2	13.0	20 45.4	09.2	54.3
	16	60 23.0	19.8	34 06.2	12.9	20 54.6	09.2	54.3
	17	75 23.1	20.6	48 38.1	12.9	21 03.8	09.0	54.3
	18	90 23.2	N12 21.5	63 10.0	12.8	N21 12.8	09.0	54.3
	19	105 23.3	22.3	77 41.8	12.7	21 21.8	08.8	54.4
	20	120 23.5	23.1	92 13.5	12.6	21 30.6	08.8	54.4
	21	135 23.6	. . 24.0	106 45.1	12.6	21 39.4	08.7	54.4
	22	150 23.7	24.8	121 16.7	12.5	21 48.1	08.6	54.4
	23	165 23.8	25.6	135 48.2	12.4	21 56.7	08.5	54.4
23	00	180 23.9	N12 26.5	150 19.6	12.4	N22 05.2	08.4	54.4
	01	195 24.1	27.3	164 51.0	12.3	22 13.6	08.3	54.4
	02	210 24.2	28.2	179 22.3	12.2	22 21.9	08.2	54.4
	03	225 24.3	. . 29.0	193 53.5	12.1	22 30.1	08.1	54.5
	04	240 24.4	29.8	208 24.6	12.0	22 38.2	08.0	54.5
	05	255 24.5	30.7	222 55.6	12.0	22 46.2	07.9	54.5
	06	270 24.6	N12 31.5	237 26.6	11.9	N22 54.1	07.8	54.5
	07	285 24.8	32.3	251 57.5	11.8	23 01.9	07.7	54.5
T	08	300 24.9	33.1	266 28.3	11.8	23 09.6	07.6	54.5
U	09	315 25.0	. . 34.0	280 59.1	11.7	23 17.2	07.5	54.5
E	10	330 25.1	34.8	295 29.8	11.6	23 24.7	07.4	54.5
S	11	345 25.2	35.6	310 00.4	11.5	23 32.1	07.3	54.6
D	12	0 25.3	N12 36.5	324 30.9	11.5	N23 39.4	07.2	54.6
A	13	15 25.5	37.3	339 01.4	11.4	23 46.6	07.1	54.6
Y	14	30 25.6	38.1	353 31.8	11.3	23 53.7	07.0	54.6
	15	45 25.7	. . 39.0	8 02.1	11.3	24 00.7	06.9	54.6
	16	60 25.8	39.8	22 32.4	11.1	24 07.6	06.7	54.6
	17	75 25.9	40.6	37 02.5	11.1	24 14.3	06.7	54.6
	18	90 26.0	N12 41.5	51 32.6	11.1	N24 21.0	06.5	54.7
	19	105 26.2	42.3	66 02.7	10.9	24 27.5	06.4	54.7
	20	120 26.3	43.1	80 32.6	10.9	24 33.9	06.3	54.7
	21	135 26.4	. . 43.9	95 02.5	10.8	24 40.2	06.2	54.7
	22	150 26.5	44.8	109 32.3	10.8	24 46.4	06.1	54.7
	23	165 26.6	45.6	124 02.1	10.6	24 52.5	06.0	54.7
24	00	180 26.7	N12 46.4	138 31.7	10.6	N24 58.5	05.8	54.8
	01	195 26.8	47.2	153 01.3	10.6	25 04.3	05.8	54.8
	02	210 27.0	48.1	167 30.9	10.4	25 10.1	05.6	54.8
	03	225 27.1	. . 48.9	182 00.3	10.4	25 15.7	05.5	54.8
	04	240 27.2	49.7	196 29.7	10.4	25 21.2	05.4	54.8
	05	255 27.3	50.5	210 59.1	10.2	25 26.6	05.2	54.8
	06	270 27.4	N12 51.4	225 28.3	10.2	N25 31.8	05.1	54.9
W	07	285 27.5	52.2	239 57.5	10.2	25 36.9	05.0	54.9
E	08	300 27.6	53.0	254 26.7	10.0	25 41.9	04.9	54.9
D	09	315 27.8	. . 53.8	268 55.7	10.0	25 46.8	04.8	54.9
N	10	330 27.9	54.7	283 24.7	10.0	25 51.6	04.6	54.9
E	11	345 28.0	55.5	297 53.7	09.8	25 56.2	04.5	55.0
S	12	0 28.1	N12 56.3	312 22.5	09.8	N26 00.7	04.4	55.0
D	13	15 28.2	57.1	326 51.3	09.8	26 05.1	04.3	55.0
A	14	30 28.3	57.9	341 20.1	09.7	26 09.4	04.1	55.0
Y	15	45 28.4	. . 58.8	355 48.8	09.6	26 13.5	04.0	55.0
	16	60 28.5	12 59.6	10 17.4	09.6	26 17.5	03.9	55.1
	17	75 28.6	13 00.4	24 46.0	09.5	26 21.4	03.7	55.1
	18	90 28.8	N13 01.2	39 14.5	09.4	N26 25.1	03.6	55.1
	19	105 28.9	02.0	53 42.9	09.4	26 28.7	03.5	55.1
	20	120 29.0	02.9	68 11.3	09.3	26 32.2	03.3	55.1
	21	135 29.1	. . 03.7	82 39.6	09.3	26 35.5	03.2	55.2
	22	150 29.2	04.5	97 07.9	09.2	26 38.7	03.1	55.2
	23	165 29.3	05.3	111 36.1	09.2	26 41.8	02.9	55.2
		S.D. 15.9	*d* 0.8	S.D. 14.8		14.9		15.0

Lat. °	Twilight Naut. h m	Twilight Civil h m	Sunrise h m	Moonrise 22 h m	Moonrise 23 h m	Moonrise 24 h m	Moonrise 25 h m
N 72	////	////	02 51	▭	▭	▭	▭
N 70	////	01 31	03 16	02 46	▭	▭	▭
68	////	02 13	03 34	03 31	▭	▭	▭
66	////	02 40	03 49	04 01	03 32	▭	▭
64	01 19	03 01	04 01	04 23	04 13	03 53	▭
62	01 57	03 18	04 12	04 41	04 41	04 44	04 58
60	02 22	03 32	04 21	04 56	05 03	05 16	05 42
N 58	02 42	03 43	04 28	05 09	05 21	05 40	06 12
56	02 58	03 53	04 35	05 20	05 36	05 59	06 35
54	03 11	04 02	04 41	05 30	05 49	06 15	06 53
52	03 22	04 09	04 47	05 39	06 00	06 29	07 09
50	03 32	04 16	04 52	05 47	06 10	06 41	07 23
45	03 52	04 31	05 02	06 04	06 32	07 06	07 50
N 40	04 08	04 42	05 11	06 18	06 49	07 27	08 12
35	04 20	04 52	05 18	06 29	07 04	07 44	08 31
30	04 31	05 00	05 25	06 40	07 16	07 58	08 46
20	04 47	05 13	05 36	06 58	07 38	08 23	09 13
N 10	04 59	05 24	05 46	07 13	07 57	08 45	09 36
0	05 09	05 34	05 55	07 28	08 15	09 05	09 57
S 10	05 18	05 42	06 04	07 43	08 33	09 25	10 19
20	05 25	05 51	06 13	07 59	08 52	09 47	10 42
30	05 31	05 59	06 24	08 17	09 15	10 12	11 09
35	05 34	06 04	06 30	08 28	09 28	10 27	11 25
40	05 37	06 09	06 37	08 41	09 43	10 45	11 43
45	05 40	06 15	06 45	08 55	10 01	11 06	12 06
S 50	05 43	06 21	06 54	09 14	10 24	11 32	12 34
52	05 44	06 24	06 59	09 22	10 35	11 45	12 48
54	05 45	06 27	07 04	09 32	10 48	12 00	13 05
56	05 47	06 30	07 09	09 43	11 02	12 18	13 24
58	05 48	06 34	07 15	09 55	11 19	12 39	13 48
S 60	05 49	06 38	07 22	10 10	11 39	13 07	14 20

Lat. °	Sunset h m	Twilight Civil h m	Twilight Naut. h m	Moonset 22 h m	Moonset 23 h m	Moonset 24 h m	Moonset 25 h m
N 72	21 10	////	////	▭	▭	▭	▭
N 70	20 45	22 35	////	▭	▭	▭	▭
68	20 25	21 49	////	▭	▭	▭	▭
66	20 10	21 20	////	24 21	00 21	▭	▭
64	19 57	20 58	22 46	23 41	25 43	01 43	▭
62	19 47	20 41	22 05	23 13	24 52	00 52	02 27
60	19 38	20 27	21 38	22 52	24 21	00 21	01 42
N 58	19 30	20 15	21 18	22 35	23 57	25 13	01 13
56	19 23	20 05	21 01	22 21	23 38	24 50	00 50
54	19 17	19 56	20 48	22 08	23 23	24 32	00 32
52	19 11	19 49	20 36	21 57	23 09	24 16	00 16
50	19 06	19 42	20 26	21 47	22 57	24 03	00 03
45	18 55	19 27	20 06	21 27	22 32	23 35	24 32
N 40	18 46	19 15	19 50	21 11	22 13	23 13	24 09
35	18 39	19 05	19 37	20 57	21 57	22 55	23 51
30	18 32	18 57	19 27	20 45	21 42	22 40	23 35
20	18 21	18 43	19 10	20 25	21 19	22 14	23 08
N 10	18 11	18 32	18 58	20 07	20 58	21 51	22 45
0	18 02	18 23	18 47	19 50	20 39	21 30	22 24
S 10	17 53	18 14	18 39	19 34	20 20	21 09	22 02
20	17 43	18 06	18 32	19 16	19 59	20 46	21 39
30	17 32	17 57	18 25	18 56	19 35	20 20	21 12
35	17 26	17 52	18 22	18 45	19 22	20 05	20 56
40	17 19	17 47	18 19	18 31	19 06	19 47	20 37
45	17 11	17 41	18 16	18 16	18 47	19 26	20 15
S 50	17 02	17 35	18 13	17 56	18 23	18 59	19 46
52	16 57	17 32	18 12	17 47	18 12	18 45	19 32
54	16 52	17 29	18 10	17 37	17 59	18 30	19 16
56	16 47	17 26	18 09	17 26	17 44	18 12	18 56
58	16 41	17 22	18 08	17 12	17 27	17 51	18 32
S 60	16 34	17 18	18 07	16 57	17 06	17 23	18 00

Day	SUN Eqn. of Time 00ʰ m s	SUN Eqn. of Time 12ʰ m s	SUN Mer. Pass. h m	MOON Mer. Pass. Upper h m	MOON Mer. Pass. Lower h m	MOON Age d	MOON Phase
22	01 24	01 30	11 59	13 39	01 17	02	◐
23	01 35	01 41	11 58	14 27	02 03	03	
24	01 47	01 52	11 58	15 17	02 52	04	

CONVERSION OF ARC TO TIME

0°–59°		60°–119°		120°–179°		180°–239°		240°–299°		300°–359°			0′·00	0′·25	0′·50	0′·75
°	h m	°	h m	°	h m	°	h m	°	h m	°	h m	′	m s	m s	m s	m s
0	0 00	60	4 00	120	8 00	180	12 00	240	16 00	300	20 00	0	0 00	0 01	0 02	0 03
1	0 04	61	4 04	121	8 04	181	12 04	241	16 04	301	20 04	1	0 04	0 05	0 06	0 07
2	0 08	62	4 08	122	8 08	182	12 08	242	16 08	302	20 08	2	0 08	0 09	0 10	0 11
3	0 12	63	4 12	123	8 12	183	12 12	243	16 12	303	20 12	3	0 12	0 13	0 14	0 15
4	0 16	64	4 16	124	8 16	184	12 16	244	16 16	304	20 16	4	0 16	0 17	0 18	0 19
5	0 20	65	4 20	125	8 20	185	12 20	245	16 20	305	20 20	5	0 20	0 21	0 22	0 23
6	0 24	66	4 24	126	8 24	186	12 24	246	16 24	306	20 24	6	0 24	0 25	0 26	0 27
7	0 28	67	4 28	127	8 28	187	12 28	247	16 28	307	20 28	7	0 28	0 29	0 30	0 31
8	0 32	68	4 32	128	8 32	188	12 32	248	16 32	308	20 32	8	0 32	0 33	0 34	0 35
9	0 36	69	4 36	129	8 36	189	12 36	249	16 36	309	20 36	9	0 36	0 37	0 38	0 39
10	0 40	70	4 40	130	8 40	190	12 40	250	16 40	310	20 40	10	0 40	0 41	0 42	0 43
11	0 44	71	4 44	131	8 44	191	12 44	251	16 44	311	20 44	11	0 44	0 45	0 46	0 47
12	0 48	72	4 48	132	8 48	192	12 48	252	16 48	312	20 48	12	0 48	0 49	0 50	0 51
13	0 52	73	4 52	133	8 52	193	12 52	253	16 52	313	20 52	13	0 52	0 53	0 54	0 55
14	0 56	74	4 56	134	8 56	194	12 56	254	16 56	314	20 56	14	0 56	0 57	0 58	0 59
15	1 00	75	5 00	135	9 00	195	13 00	255	17 00	315	21 00	15	1 00	1 01	1 02	1 03
16	1 04	76	5 04	136	9 04	196	13 04	256	17 04	316	21 04	16	1 04	1 05	1 06	1 07
17	1 08	77	5 08	137	9 08	197	13 08	257	17 08	317	21 08	17	1 08	1 09	1 10	1 11
18	1 12	78	5 12	138	9 12	198	13 12	258	17 12	318	21 12	18	1 12	1 13	1 14	1 15
19	1 16	79	5 16	139	9 16	199	13 16	259	17 16	319	21 16	19	1 16	1 17	1 18	1 19
20	1 20	80	5 20	140	9 20	200	13 20	260	17 20	320	21 20	20	1 20	1 21	1 22	1 23
21	1 24	81	5 24	141	9 24	201	13 24	261	17 24	321	21 24	21	1 24	1 25	1 26	1 27
22	1 28	82	5 28	142	9 28	202	13 28	262	17 28	322	21 28	22	1 28	1 29	1 30	1 31
23	1 32	83	5 32	143	9 32	203	13 32	263	17 32	323	21 32	23	1 32	1 33	1 34	1 35
24	1 36	84	5 36	144	9 36	204	13 36	264	17 36.	324	21 36	24	1 36	1 37	1 38	1 39
25	1 40	85	5 40	145	9 40	205	13 40	265	17 40	325	21 40	25	1 40	1 41	1 42	1 43
26	1 44	86	5 44	146	9 44	206	13 44	266	17 44	326	21 44	26	1 44	1 45	1 46	1 47
27	1 48	87	5 48	147	9 48	207	13 48	267	17 48	327	21 48	27	1 48	1 49	1 50	1 51
28	1 52	88	5 52	148	9 52	208	13 52	268	17 52	328	21 52	28	1 52	1 53	1 54	1 55
29	1 56	89	5 56	149	9 56	209	13 56	269	17 56	329	21 56	29	1 56	1 57	1 58	1 59
30	2 00	90	6 00	150	10 00	210	14 00	270	18 00	330	22 00	30	2 00	2 01	2 02	2 03
31	2 04	91	6 04	151	10 04	211	14 04	271	18 04	331	22 04	31	2 04	2 05	2 06	2 07
32	2 08	92	6 08	152	10 08	212	14 08	272	18 08	332	22 08	32	2 08	2 09	2 10	2 11
33	2 12	93	6 12	153	10 12	213	14 12	273	18 12	333	22 12	33	2 12	2 13	2 14	2 15
34	2 16	94	6 16	154	10 16	214	14 16	274	18 16	334	22 16	34	2 16	2 17	2 18	2 19
35	2 20	95	6 20	155	10 20	215	14 20	275	18 20	335	22 20	35	2 20	2 21	2 22	2 23
36	2 24	96	6 24	156	10 24	216	14 24	276	18 24	336	22 24	36	2 24	2 25	2 26	2 27
37	2 28	97	6 28	157	10 28	217	14 28	277	18 28	337	22 28	37	2 28	2 29	2 30	2 31
38	2 32	98	6 32	158	10 32	218	14 32	278	18 32	338	22 32	38	2 32	2 33	2 34	2 35
39	2 36	99	6 36	159	10 36	219	14 36	279	18 36	339	22 36	39	2 36	2 37	2 38	2 39
40	2 40	100	6 40	160	10 40	220	14 40	280	18 40	340	22 40	40	2 40	2 41	2 42	2 43
41	2 44	101	6 44	161	10 44	221	14 44	281	18 44	341	22 44	41	2 44	2 45	2 46	2 47
42	2 48	102	6 48	162	10 48	222	14 48	282	18 48	342	22 48	42	2 48	2 49	2 50	2 51
43	2 52	103	6 52	163	10 52	223	14 52	283	18 52	343	22 52	43	2 52	2 53	2 54	2 55
44	2 56	104	6 56	164	10 56	224	14 56	284	18 56	344	22 56	44	2 56	2 57	2 58	2 59
45	3 00	105	7 00	165	11 00	225	15 00	285	19 00	345	23 00	45	3 00	3 01	3 02	3 03
46	3 04	106	7 04	166	11 04	226	15 04	286	19 04	346	23 04	46	3 04	3 05	3 06	3 07
47	3 08	107	7 08	167	11 08	227	15 08	287	19 08	347	23 08	47	3 08	3 09	3 10	3 11
48	3 12	108	7 12	168	11 12	228	15 12	288	19 12	348	23 12	48	3 12	3 13	3 14	3 15
49	3 16	109	7 16	169	11 16	229	15 16	289	19 16	349	23 16	49	3 16	3 17	3 18	3 19
50	3 20	110	7 20	170	11 20	230	15 20	290	19 20	350	23 20	50	3 20	3 21	3 22	3 23
51	3 24	111	7 24	171	11 24	231	15 24	291	19 24	351	23 24	51	3 24	3 25	3 26	3 27
52	3 28	112	7 28	172	11 28	232	15 28	292	19 28	352	23 28	52	3 28	3 29	3 30	3 31
53	3 32	113	7 32	173	11 32	233	15 32	293	19 32	353	23 32	53	3 32	3 33	3 34	3 35
54	3 36	114	7 36	174	11 36	234	15 36	294	19 36	354	23 36	54	3 36	3 37	3 38	3 39
55	3 40	115	7 40	175	11 40	235	15 40	295	19 40	355	23 40	55	3 40	3 41	3 42	3 43
56	3 44	116	7 44	176	11 44	236	15 44	296	19 44	356	23 44	56	3 44	3 45	3 46	3 47
57	3 48	117	7 48	177	11 48	237	15 48	297	19 48	357	23 48	57	3 48	3 49	3 50	3 51
58	3 52	118	7 52	178	11 52	238	15 52	298	19 52	358	23 52	58	3 52	3 53	3 54	3 55
59	3 56	119	7 56	179	11 56	239	15 56	299	19 56	359	23 56	59	3 56	3 57	3 58	3 59

The above table is for converting expressions in arc to their equivalent in time ; its main use in this Almanac is for the conversion of longitude for application to L.M.T. (*added* if *west*, *subtracted* if *east*) to give G.M.T. or vice versa, particularly in the case of sunrise, sunset, etc.

TABLES FOR INTERPOLATING SUNRISE, MOONRISE, ETC.

TABLE I—FOR LATITUDE

Tabular Interval 10°	5°	2°	Difference between the times for consecutive latitudes 5m	10m	15m	20m	25m	30m	35m	40m	45m	50m	55m	60m	1h 05m	1h 10m	1h 15m	1h 20m
° ′	° ′	° ′	m	m	m	m	m	m	m	m	m	m	m	m	h m	h m	h m	h m
0 30	0 15	0 06	0	0	1	1	1	1	1	2	2	2	2	2	0 02	0 02	0 02	0 02
1 00	0 30	0 12	0	1	1	2	2	3	3	3	4	4	4	5	05	05	05	05
1 30	0 45	0 18	1	1	2	3	3	4	4	5	5	6	7	7	07	07	07	07
2 00	1 00	0 24	1	2	3	4	5	5	6	7	7	8	9	10	10	10	10	10
2 30	1 15	0 30	1	2	4	5	6	7	8	9	9	10	11	12	12	13	13	13
3 00	1 30	0 36	1	3	4	6	7	8	9	10	11	12	13	14	0 15	0 15	0 16	0 16
3 30	1 45	0 42	2	3	5	7	8	10	11	12	13	14	16	17	18	18	19	19
4 00	2 00	0 48	2	4	6	8	9	11	13	14	15	16	18	19	20	21	22	22
4 30	2 15	0 54	2	4	7	9	11	13	15	16	18	19	21	22	23	24	25	26
5 00	2 30	1 00	2	5	7	10	12	14	16	18	20	22	23	25	26	27	28	29
5 30	2 45	1 06	3	5	8	11	13	16	18	20	22	24	26	28	0 29	0 30	0 31	0 32
6 00	3 00	1 12	3	6	9	12	14	17	20	22	24	26	29	31	32	33	34	36
6 30	3 15	1 18	3	6	10	13	16	19	22	24	26	29	31	34	36	37	38	40
7 00	3 30	1 24	3	7	10	14	17	20	23	26	29	31	34	37	39	41	42	44
7 30	3 45	1 30	4	7	11	15	18	22	25	28	31	34	37	40	43	44	46	48
8 00	4 00	1 36	4	8	12	16	20	23	27	30	34	37	41	44	0 47	0 48	0 51	0 53
8 30	4 15	1 42	4	8	13	17	21	25	29	33	36	40	44	48	0 51	0 53	0 56	0 58
9 00	4 30	1 48	4	9	13	18	22	27	31	35	39	43	47	52	0 55	0 58	1 01	1 04
9 30	4 45	1 54	5	9	14	19	24	28	33	38	42	47	51	56	1 00	1 04	1 08	1 12
10 00	5 00	2 00	5	10	15	20	25	30	35	40	45	50	55	60	1 05	1 10	1 15	1 20

Table I is for interpolating the L.M.T. of sunrise, twilight, moonrise, etc., for latitude. It is to be entered, in the appropriate column on the left, with the difference between true latitude and the nearest tabular latitude which is *less* than the true latitude; and with the argument at the top which is the nearest value of the difference between the times for the tabular latitude and the next higher one; the correction so obtained is applied to the time for the tabular latitude; the sign of the correction can be seen by inspection. It is to be noted that the interpolation is not linear, so that when using this table it is essential to take out the tabular phenomenon for the latitude *less* than the true latitude.

TABLE II—FOR LONGITUDE

Long. East or West	Difference between the times for given date and preceding date (for east longitude) or for given date and following date (for west longitude) 10m	20m	30m	40m	50m	60m	1h + 10m	1h + 20m	1h + 30m	1h + 40m	1h + 50m	1h + 60m	2h 10m	2h 20m	2h 30m	2h 40m	2h 50m	3h 00m
°	m	m	m	m	m	m	m	m	m	m	m	m	h m	h m	h m	h m	h m	h m
0	0	0	0	0	0	0	0	0	0	0	0	0	0 00	0 00	0 00	0 00	0 00	0 00
10	0	1	1	1	1	2	2	2	2	3	3	3	04	04	04	04	05	05
20	1	1	2	2	3	3	4	4	5	6	6	7	07	08	08	09	09	10
30	1	2	2	3	4	5	6	7	7	8	9	10	11	12	12	13	14	15
40	1	2	3	4	6	7	8	9	10	11	12	13	14	16	17	18	19	20
50	1	3	4	6	7	8	10	11	12	14	15	17	0 18	0 19	0 21	0 22	0 24	0 25
60	2	3	5	7	8	10	12	13	15	17	18	20	22	23	25	27	28	30
70	2	4	6	8	10	12	14	16	17	19	21	23	25	27	29	31	33	35
80	2	4	7	9	11	13	16	18	20	22	24	27	29	31	33	36	38	40
90	2	5	7	10	12	15	17	20	22	25	27	30	32	35	37	40	42	45
100	3	6	8	11	14	17	19	22	25	28	31	33	0 36	0 39	0 42	0 44	0 47	0 50
110	3	6	9	12	15	18	21	24	27	31	34	37	40	43	46	49	0 52	0 55
120	3	7	10	13	17	20	23	27	30	33	37	40	43	47	50	53	0 57	1 00
130	4	7	11	14	18	22	25	29	32	36	40	43	47	51	54	0 58	1 01	1 05
140	4	8	12	16	19	23	27	31	35	39	43	47	51	54	0 58	1 02	1 06	1 10
150	4	8	13	17	21	25	29	33	38	42	46	50	0 54	0 58	1 03	1 07	1 11	1 15
160	4	9	13	18	22	27	31	36	40	44	49	53	0 58	1 02	1 07	1 11	1 16	1 20
170	5	9	14	19	24	28	33	38	42	47	52	57	1 01	1 06	1 11	1 16	1 20	1 25
180	5	10	15	20	25	30	35	40	45	50	55	60	1 05	1 10	1 15	1 20	1 25	1 30

Table II is for interpolating the L.M.T. of moonrise, moonset and the Moon's meridian passage for longitude. It is entered with longitude and with the difference between the times for the given date and for the preceding date (in east longitudes) or following date (in west longitudes). The correction is normally *added* for west longitudes and *subtracted* for east longitudes, but if, as occasionally happens, the times become earlier each day instead of later, the signs of the corrections must be reversed.

LHA ♈	Hc	Zn	Hc	Zn	Hc	Zn	Hc	Zn	Hc	Zn	Hc	Zn	Hc	Zn
	•CAPELLA		ALDEBARAN		•Diphda		FOMALHAUT		ALTAIR		•VEGA		Kochab	
0	32 04	054	26 36	088	34 59	168	22 38	195	27 21	261	27 54	299	23 49	348
1	32 43	054	27 24	089	35 09	169	22 26	196	26 33	261	27 12	300	23 40	349
2	33 23	054	28 13	090	35 18	170	22 12	197	25 45	262	26 30	300	23 30	349
3	34 02	054	29 02	090	35 26	171	21 58	198	24 57	263	25 48	301	23 21	349
4	34 42	055	29 50	091	35 34	172	21 43	198	24 09	263	25 06	301	23 12	349
5	35 21	055	30 39	091	35 40	173	21 27	199	23 21	264	24 25	301	23 03	350
6	36 01	055	31 27	092	35 45	174	21 11	200	22 33	264	23 43	302	22 54	350
7	36 41	055	32 16	093	35 49	176	20 53	201	21 44	265	23 02	302	22 46	350
8	37 21	056	33 04	093	35 52	177	20 36	202	20 56	266	22 21	303	22 37	350
9	38 01	056	33 53	094	35 54	178	20 17	203	20 07	266	21 40	303	22 29	351
10	38 41	056	34 41	094	35 56	179	19 58	204	19 19	267	21 00	303	22 22	351
11	39 22	056	35 29	095	35 56	180	19 38	205	18 30	268	20 19	304	22 14	351
12	40 02	057	36 18	096	35 55	182	19 18	205	17 42	268	19 39	304	22 07	351
13	40 43	057	37 06	096	35 53	183	18 56	206	16 53	269	18 59	305	21 59	352
14	41 24	057	37 54	097	35 50	184	18 35	207	16 05	269	18 19	305	21 52	352
	•CAPELLA		BETELGEUSE		RIGEL		•Diphda		Enif		•DENEB		Kochab	
15	42 04	057	17 36	094	15 54	113	35 47	185	38 26	252	41 19	302	21 46	352
16	42 45	057	18 25	094	16 38	114	35 42	186	37 40	253	40 38	302	21 39	352
17	43 26	058	19 13	095	17 22	114	35 36	187	36 54	254	39 56	302	21 33	353
18	44 07	058	20 01	095	18 07	115	35 30	189	36 07	254	39 16	303	21 27	353
19	44 48	058	20 50	096	18 50	116	35 22	190	35 20	255	38 35	303	21 21	353
20	45 29	058	21 38	097	19 34	116	35 13	191	34 33	256	37 54	303	21 16	354
21	46 11	058	22 26	097	20 17	117	35 04	192	33 46	257	37 13	303	21 10	354
22	46 52	058	23 14	098	21 00	118	34 53	193	32 59	257	36 33	304	21 05	354
23	47 33	058	24 02	099	21 43	119	34 42	194	32 11	258	35 53	304	21 00	354
24	48 15	059	24 50	099	22 26	119	34 29	195	31 24	259	35 12	304	20 56	355
25	48 56	059	25 38	100	23 08	120	34 16	196	30 36	259	34 32	304	20 51	355
26	49 38	059	26 26	101	23 49	121	34 02	198	29 48	260	33 52	305	20 47	355
27	50 19	059	27 13	101	24 31	122	33 46	199	29 00	261	33 12	305	20 43	355
28	51 01	059	28 01	102	25 12	122	33 30	200	28 12	261	32 33	305	20 39	356
29	51 42	059	28 48	103	25 53	123	33 14	201	27 24	262	31 53	306	20 36	356
	•CAPELLA		BETELGEUSE		RIGEL		•Diphda		Alpheratz		•DENEB		Kochab	
30	52 24	059	29 36	103	26 33	124	32 56	202	65 24	262	31 14	306	20 33	356
31	53 06	059	30 23	104	27 13	125	32 37	203	64 35	262	30 34	306	20 30	357
32	53 47	059	31 10	105	27 53	126	32 18	204	63 47	263	29 55	306	20 27	357
33	54 29	059	31 57	105	28 32	127	31 58	205	62 59	264	29 16	307	20 25	357
34	55 11	059	32 44	106	29 11	127	31 37	206	62 11	265	28 38	307	20 22	357
35	55 52	059	33 30	107	29 49	128	31 15	207	61 22	265	27 59	307	20 20	358
36	56 34	059	34 16	108	30 27	129	30 52	208	60 34	266	27 20	308	20 18	358
37	57 16	059	35 03	108	31 04	130	30 29	209	59 46	267	26 42	308	20 17	358
38	57 57	059	35 49	109	31 41	131	30 05	210	58 57	267	26 04	308	20 16	359
39	58 39	059	36 34	110	32 18	132	29 40	211	58 09	268	25 26	309	20 15	359
40	59 21	059	37 20	111	32 53	133	29 14	212	57 20	269	24 48	309	20 14	359
41	60 02	059	38 05	112	33 29	134	28 48	213	56 31	269	24 11	309	20 13	000
42	60 43	059	38 50	112	34 03	135	28 21	214	55 43	270	23 33	310	20 13	000
43	61 25	058	39 35	113	34 38	136	27 54	215	54 54	270	22 56	310	20 13	000

LAT 36°N

LHA ♈	Hc	Zn	Hc	Zn	Hc	Zn	Hc	Zn	Hc	Zn	Hc	Zn	Hc	Zn
	•Dubhe		REGULUS		PROCYON		•SIRIUS		RIGEL		ALDEBARAN		•Mirfak	
90	37 46	035	29 42	097	51 50	138	36 16	167	44 28	196	62 51	230	58 37	308
91	38 14	035	30 30	097	52 22	139	36 26	168	44 14	197	62 14	231	57 59	308
92	38 42	035	31 18	098	52 54	141	36 36	169	43 59	199	61 36	233	57 21	308
93	39 11	036	32 06	099	53 24	142	36 45	170	43 43	200	60 57	234	56 42	308
94	39 39	036	32 54	099	53 53	143	36 53	171	43 25	201	60 17	236	56 04	308
95	40 07	036	33 42	100	54 22	145	36 59	173	43 07	203	59 37	237	55 25	308
96	40 35	036	34 29	101	54 49	146	37 05	174	42 48	204	58 56	238	54 47	307
97	41 04	036	35 17	101	55 15	148	37 10	175	42 28	205	58 14	240	54 08	307
98	41 32	036	36 05	102	55 40	150	37 13	176	42 07	207	57 32	241	53 30	307
99	42 00	036	36 52	103	56 04	151	37 16	177	41 44	208	56 49	242	52 51	307
100	42 29	036	37 39	104	56 27	153	37 18	179	41 21	209	56 06	243	52 12	307
101	42 57	036	38 26	104	56 49	155	37 18	180	40 57	210	55 23	244	51 34	307
102	43 25	036	39 13	105	57 09	156	37 18	181	40 33	211	54 39	245	50 55	307
103	43 54	036	40 00	106	57 27	158	37 17	182	40 07	213	53 55	246	50 16	307
104	44 22	036	40 47	107	57 45	160	37 14	183	39 40	214	53 10	247	49 38	307
	Kochab		•Denebola		REGULUS		•SIRIUS		RIGEL		ALDEBARAN		•CAPELLA	
105	27 36	016	22 56	088	41 33	107	37 11	185	39 13	215	52 25	248	68 03	305
106	27 50	016	23 44	089	42 19	108	37 06	186	38 45	216	51 40	249	67 23	305
107	28 03	016	24 33	089	43 05	109	37 01	187	38 16	217	50 54	250	66 43	304
108	28 17	016	25 21	090	43 51	110	36 54	188	37 47	218	50 08	251	66 03	304
109	28 30	016	26 10	090	44 37	111	36 47	189	37 16	219	49 22	252	65 22	303
110	28 44	017	26 59	091	45 22	112	36 38	191	36 45	220	48 36	253	64 42	303
111	28 58	017	27 47	092	46 07	113	36 29	192	36 13	221	47 50	254	64 01	303
112	29 12	017	28 36	092	46 52	113	36 19	193	35 41	222	47 03	255	63 20	302
113	29 26	017	29 24	093	47 36	114	36 07	194	35 08	223	46 16	255	62 39	302
114	29 41	017	30 13	094	48 20	115	35 55	195	34 34	224	45 29	256	61 57	302
115	29 55	017	31 01	094	49 04	116	35 42	196	34 00	225	44 42	257	61 16	302
116	30 10	018	31 49	095	49 47	117	35 27	198	33 25	226	43 54	258	60 35	301
117	30 24	018	32 38	095	50 30	118	35 12	199	32 50	227	43 07	258	59 53	301
118	30 39	018	33 26	096	51 12	119	34 56	200	32 14	228	42 19	259	59 12	301
119	30 54	018	34 14	097	51 54	121	34 39	201	31 38	229	41 32	260	58 30	301
	•Kochab		Denebola		•REGULUS		SIRIUS		RIGEL		•ALDEBARAN		CAPELLA	
120	31 09	018	35 02	097	52 36	122	34 22	202	31 01	230	40 44	261	57 49	301
121	31 24	018	35 51	098	53 17	123	34 03	203	30 23	231	39 56	261	57 07	301
122	31 39	018	36 39	099	53 57	124	33 43	204	29 46	232	39 08	262	56 25	301
123	31 55	018	37 27	099	54 37	125	33 23	205	29 07	233	38 20	263	55 44	301
124	32 10	019	38 14	100	55 17	127	33 02	206	28 28	233	37 31	263	55 02	301
125	32 26	019	39 02	101	55 55	128	32 40	207	27 49	234	36 43	264	54 20	301
126	32 41	019	39 50	101	56 33	129	32 17	208	27 10	235	35 55	265	53 39	301
127	32 57	019	40 37	102	57 10	131	31 54	209	26 29	236	35 07	265	52 57	301
128	33 12	019	41 25	103	57 47	132	31 29	211	25 49	237	34 18	266	52 15	301
129	33 28	019	42 12	104	58 23	133	31 04	212	25 08	238	33 30	267	51 34	301
130	33 44	019	42 59	104	58 57	135	30 39	213	24 27	238	32 41	267	50 52	301
131	34 00	019	43 46	105	59 31	137	30 12	214	23 46	239	31 53	268	50 10	301
132	34 16	019	44 33	106	60 04	138	29 45	214	23 04	240	31 04	268	49 29	301
133	34 32	019	45 19	107	60 36	140	29 17	215	22 22	241	30 16	269	48 47	301

45	18 51 025	30 31 076	15 15 124	35 44 138	26 57 217	53 17 272	21 42 311
46	19 12 026	31 18 076	15 55 125	36 16 139	26 27 218	52 29 272	21 06 311
47	19 33 026	32 05 077	16 35 126	36 48 140	25 57 219	51 40 273	20 30 312
48	19 55 026	32 52 077	17 14 127	37 19 141	25 26 220	50 52 273	19 53 312
49	20 17 027	33 40 078	17 52 127	37 49 142	24 55 221	50 03 274	19 18 312
50	20 39 027	34 27 078	18 31 128	38 19 143	24 23 221	49 15 274	18 42 313
51	21 01 027	35 15 079	19 09 129	38 48 144	23 51 222	48 27 275	18 06 313
52	21 23 028	36 02 079	19 46 130	39 16 145	23 18 223	47 38 275	17 31 314
53	21 46 028	36 50 079	20 24 130	39 43 146	22 45 224	46 50 276	16 56 314
54	22 09 028	37 38 080	21 00 131	40 09 148	22 11 225	46 02 276	16 21 315
55	22 32 029	38 26 080	21 37 132	40 35 149	21 36 226	45 13 277	15 47 315
56	22 55 029	39 14 081	22 12 133	41 00 150	21 01 226	44 25 277	15 13 315
57	23 19 029	40 02 081	22 48 134	41 24 151	20 26 227	43 37 278	14 39 316
58	23 42 029	40 50 082	23 23 134	41 46 152	19 50 228	42 49 278	14 05 316
59	24 06 030	41 38 082	23 57 135	42 09 154	19 14 229	42 01 279	13 32 317
	•Dubhe	POLLUX	PROCYON	•SIRIUS	RIGEL	•Hamal	Schedar
60	24 30 030	42 26 083	31 21 108	24 31 136	42 30 155	62 26 251	50 58 318
61	24 55 030	43 14 083	32 07 109	25 04 137	42 50 156	61 40 252	50 25 318
62	25 19 030	44 02 084	32 53 110	25 37 138	43 09 157	60 53 253	49 52 317
63	25 44 031	44 51 084	33 39 110	26 09 139	43 27 159	60 07 254	49 19 317
64	26 09 031	45 39 085	34 24 111	26 41 140	43 44 160	59 20 255	48 46 317
65	26 34 031	46 27 085	35 09 112	27 12 141	44 00 161	58 33 256	48 13 317
66	26 59 031	47 16 086	35 54 113	27 43 142	44 15 163	57 46 257	47 40 317
67	27 25 032	48 04 086	36 39 114	28 12 142	44 29 164	56 59 257	47 07 317
68	27 50 032	48 53 087	37 23 114	28 42 143	44 42 165	56 11 258	46 34 317
69	28 16 032	49 41 088	38 07 115	29 10 144	44 54 167	55 24 259	46 01 317
70	28 42 032	50 30 088	38 51 116	29 38 145	45 04 168	54 36 260	45 28 317
71	29 08 033	51 18 089	39 34 117	30 06 146	45 14 169	53 48 261	44 55 317
72	29 34 033	52 07 089	40 17 118	30 32 147	45 22 171	53 00 261	44 22 317
73	30 01 033	52 55 090	41 00 119	30 58 148	45 29 172	52 12 262	43 48 317
74	30 27 033	53 44 090	41 42 120	31 23 149	45 35 174	51 24 263	43 15 317
	•Dubhe	POLLUX	PROCYON	•SIRIUS	RIGEL	•Hamal	Schedar
75	30 54 033	54 32 091	42 24 121	31 48 150	45 40 175	50 36 264	42 42 317
76	31 21 034	55 21 092	43 06 122	32 12 151	45 43 177	49 48 264	42 09 317
77	31 47 034	56 09 092	43 47 123	32 34 152	45 46 178	48 59 265	41 36 317
78	32 14 034	56 58 093	44 28 124	32 57 153	45 47 179	48 11 266	41 03 317
79	32 42 034	57 46 094	45 08 125	33 18 154	45 47 181	47 22 266	40 30 317
80	33 09 034	58 35 094	45 48 126	33 39 156	45 45 182	46 34 267	39 57 317
81	33 36 034	59 23 095	46 27 127	33 58 157	45 43 184	45 45 268	39 24 317
82	34 04 034	60 11 096	47 05 128	34 17 158	45 39 185	44 57 268	38 51 318
83	34 31 035	61 00 096	47 43 129	34 35 159	45 35 186	44 08 269	38 19 318
84	34 59 035	61 48 097	48 21 130	34 52 160	45 28 188	43 20 269	37 46 318
85	35 26 035	62 36 098	48 57 131	35 08 161	45 21 189	42 31 270	37 13 318
86	35 54 035	63 24 099	49 33 133	35 24 162	45 13 191	41 43 271	36 41 318
87	36 22 035	64 12 100	50 09 134	35 38 163	45 03 192	40 54 271	36 09 318
88	36 50 035	65 00 100	50 43 135	35 52 164	44 53 193	40 06 272	35 36 318
89	37 18 035	65 47 101	51 17 136	36 04 166	44 41 195	39 17 272	35 04 318

135	35 04 019	20 03 080	12 03 113	61 36 143	28 20 217	38 58 247	47 25 302
136	35 20 019	20 51 081	12 47 114	62 05 145	27 50 218	38 13 248	46 43 302
137	35 36 019	21 39 081	13 31 115	62 32 147	27 19 219	37 28 249	46 02 302
138	35 52 020	22 27 082	14 15 115	62 58 149	26 49 220	36 43 250	45 21 302
139	36 09 020	23 15 082	14 59 116	63 22 151	26 17 221	35 57 251	44 40 302
140	36 25 020	24 04 083	15 42 117	63 45 153	25 45 222	35 11 252	43 58 302
141	36 41 020	24 52 084	16 26 118	64 06 155	25 12 223	34 25 252	43 17 302
142	36 58 020	25 40 084	17 08 118	64 26 157	24 39 224	33 38 253	42 37 303
143	37 14 020	26 28 085	17 51 119	64 44 159	24 05 224	32 52 254	41 56 303
144	37 30 020	27 17 085	18 33 120	65 00 161	23 31 225	32 05 254	41 15 303
145	37 46 020	28 05 086	19 15 120	65 15 164	22 56 226	31 18 255	40 34 303
146	38 03 020	28 53 086	19 57 121	65 27 166	22 21 227	30 31 256	39 54 303
147	38 19 020	29 42 087	20 39 122	65 38 168	21 45 228	29 44 257	39 13 304
148	38 35 020	30 30 087	21 20 123	65 47 171	21 09 229	28 57 257	38 33 304
149	38 52 020	31 19 088	22 00 123	65 54 173	20 32 229	28 09 258	37 53 304
	•Kochab	ARCTURUS	•SPICA	REGULUS	PROCYON	•POLLUX	CAPELLA
150	39 08 020	32 07 089	22 41 124	65 59 175	45 18 235	60 21 264	37 13 304
151	39 24 020	32 56 089	23 21 125	66 02 178	44 38 236	59 33 265	36 33 305
152	39 40 019	33 44 090	24 00 126	66 02 180	43 58 237	58 44 266	35 53 305
153	39 57 019	34 33 090	24 40 127	66 01 183	43 17 238	57 56 266	35 13 305
154	40 13 019	35 22 091	25 18 127	65 58 185	42 35 239	57 07 267	34 33 305
155	40 29 019	36 10 091	25 57 128	65 53 187	41 53 240	56 19 268	33 54 306
156	40 45 019	36 59 092	26 35 129	65 45 190	41 11 241	55 30 268	33 14 306
157	41 01 019	37 47 093	27 12 130	65 36 192	40 28 242	54 42 269	32 35 306
158	41 17 019	38 36 093	27 49 131	65 25 194	39 45 243	53 53 269	31 56 306
159	41 33 019	39 24 094	28 26 132	65 12 197	39 02 244	53 05 270	31 17 307
160	41 48 019	40 12 095	29 02 132	64 57 199	38 18 245	52 16 271	30 38 307
161	42 04 019	41 01 095	29 37 133	64 40 201	37 34 245	51 28 271	30 00 307
162	42 20 019	41 49 096	30 12 134	64 22 203	36 50 246	50 39 272	29 21 308
163	42 35 019	42 37 097	30 47 135	64 02 205	36 06 247	49 51 272	28 43 308
164	42 51 018	43 26 097	31 21 136	63 40 208	35 21 248	49 02 273	28 05 308
	•Kochab	ARCTURUS	•SPICA	REGULUS	PROCYON	•POLLUX	CAPELLA
165	43 06 018	44 14 098	31 54 137	63 17 210	34 36 249	48 14 273	27 27 309
166	43 21 018	45 02 099	32 27 138	62 52 212	33 50 249	47 25 274	26 49 309
167	43 36 018	45 50 099	32 59 139	62 26 213	33 05 250	46 37 274	26 11 309
168	43 51 018	46 37 100	33 31 140	61 59 215	32 19 251	45 49 275	25 34 310
169	44 06 018	47 25 101	34 02 141	61 30 217	31 33 252	45 00 275	24 56 310
170	44 21 018	48 13 102	34 32 142	61 00 219	30 47 252	44 12 276	24 19 310
171	44 36 017	49 00 102	35 01 143	60 29 221	30 00 253	43 24 277	23 42 311
172	44 50 017	49 48 103	35 30 144	59 57 222	29 14 254	42 35 277	23 06 311
173	45 04 017	50 35 104	35 58 145	59 24 224	28 27 255	41 47 278	22 29 311
174	45 19 017	51 22 105	36 26 146	58 50 225	27 40 255	40 59 278	21 53 312
175	45 33 017	52 09 106	36 52 147	58 15 227	26 53 256	40 11 278	21 17 312
176	45 46 016	52 55 107	37 18 148	57 39 228	26 06 257	39 23 279	20 41 313
177	46 00 016	53 42 107	37 43 150	57 03 230	25 19 257	38 35 279	20 05 313
178	46 14 016	54 28 108	38 07 151	56 25 231	24 31 258	37 47 280	19 30 313
179	46 27 016	55 14 109	38 30 152	55 47 232	23 44 259	37 00 280	18 55 314

LAT 36°N

LHA ♈	Hc	Zn	Hc	Zn	Hc	Zn	Hc	Zn	Hc	Zn	Hc	Zn	Hc	Zn
	•VEGA		Alphecca		ARCTURUS		•SPICA		REGULUS→		•POLLUX		Dubhe	
180	15 34	053	43 58	086	55 59	110	38 53	153	55 08	234	36 12	281	62 39	345
181	16 12	053	44 46	086	56 45	111	39 14	154	54 29	235	35 24	281	62 27	344
182	16 52	054	45 35	087	57 30	112	39 35	155	53 49	236	34 37	282	62 13	343
183	17 31	054	46 23	087	58 15	114	39 55	157	53 08	237	33 49	282	61 59	343
184	18 10	055	47 12	088	58 59	115	40 14	158	52 27	239	33 02	283	61 44	342
185	18 50	055	48 00	089	59 43	116	40 32	159	51 45	240	32 15	283	61 29	341
186	19 30	056	48 49	089	60 26	117	40 48	160	51 03	241	31 27	284	61 12	340
187	20 10	056	49 37	090	61 09	118	41 04	162	50 21	242	30 40	284	60 56	339
188	20 51	057	50 26	090	61 52	120	41 19	163	49 38	243	29 53	285	60 38	339
189	21 31	057	51 14	091	62 34	121	41 33	164	48 54	244	29 06	285	60 20	338
190	22 12	057	52 03	092	63 15	122	41 46	165	48 11	245	28 20	286	60 02	337
191	22 53	058	52 52	092	63 56	124	41 58	167	47 26	246	27 33	286	59 42	336
192	23 34	058	53 40	093	64 36	125	42 08	168	46 42	247	26 46	287	59 23	336
193	24 16	059	54 29	093	65 15	127	42 18	169	45 57	248	26 00	287	59 03	335
194	24 57	059	55 17	094	65 53	129	42 27	171	45 12	249	25 13	287	58 42	334
	Kochab		•VEGA		Alphecca		•SPICA		REGULUS		•POLLUX		Dubhe	
195	49 30	011	25 39	059	56 05	095	42 34	172	44 27	249	24 27	288	58 21	334
196	49 40	011	26 21	060	56 54	095	42 40	173	43 41	250	23 41	288	57 59	333
197	49 49	011	27 03	060	57 42	096	42 45	175	42 56	251	22 55	289	57 37	333
198	49 57	010	27 45	061	58 30	097	42 50	176	42 10	252	22 09	289	57 15	332
199	50 06	010	28 27	061	59 18	098	42 52	177	41 23	253	21 24	290	56 52	332
200	50 14	009	29 10	061	60 06	098	42 54	179	40 37	254	20 38	290	56 29	331
201	50 22	009	29 52	062	60 54	099	42 55	180	39 50	254	19 53	291	56 05	331
202	50 29	009	30 35	062	61 42	100	42 55	181	39 03	255	19 07	291	55 41	330
203	50 36	008	31 18	062	62 30	101	42 53	183	38 16	256	18 22	292	55 17	330
204	50 43	008	32 01	063	63 18	102	42 50	184	37 29	257	17 37	292	54 52	329
205	50 50	008	32 44	063	64 05	103	42 46	185	36 42	257	16 52	293	54 28	329
206	50 56	007	33 28	063	64 52	104	42 41	187	35 55	258	16 08	293	54 03	329
207	51 02	007	34 11	064	65 39	105	42 35	188	35 07	259	15 23	294	53 37	328
208	51 07	006	34 55	064	66 26	106	42 28	189	34 19	259	14 39	294	53 12	328
209	51 13	006	35 38	064	67 13	107	42 20	191	33 32	260	13 55	295	52 46	328
	DENEB		•VEGA		Rasalhague		ANTARES		•SPICA		REGULUS		•Dubhe	
210	18 26	047	36 22	065	36 39	102	18 27	145	42 10	192	32 44	261	52 20	327
211	19 01	047	37 06	065	37 27	103	18 54	146	42 00	193	31 56	261	51 53	327
212	19 37	048	37 50	065	38 14	103	19 21	147	41 48	194	31 08	262	51 27	327
213	20 13	048	38 35	066	39 01	104	19 47	148	41 36	196	30 20	263	51 00	327
214	20 49	049	39 19	066	39 48	105	20 13	149	41 22	197	29 31	263	50 33	326
215	21 26	049	40 03	066	40 35	106	20 38	149	41 07	198	28 43	264	50 06	326
216	22 02	049	40 48	067	41 22	106	21 02	150	40 52	199	27 55	265	49 39	326
217	22 39	050	41 32	067	42 08	107	21 26	151	40 35	201	27 06	265	49 12	326
218	23 16	050	42 17	067	42 54	108	21 49	152	40 17	202	26 18	266	48 44	326
219	23 54	050	43 02	068	43 40	109	22 12	153	39 59	203	25 30	267	48 17	325
220	24 31	051	43 47	068	44 26	110	22 33	154	39 39	204	24 41	267	47 49	325
221	25 09	051	44 32	068	45 12	110	22 55	155	39 19	206	23 53	268	47 21	325
222	25 47	051	45 17	069	45 57	111	23 15	156	38 57	207	23 04	268	46 54	325
223	26 25	052	46 02	069	46 42	112	23 35	156	38 35	208	22 16	269	46 26	325
224	[illegible]	[illegible]	[illegible]	[illegible]	[illegible]	[illegible]	[illegible]	[illegible]	[illegible]	[illegible]	[illegible]	[illegible]	[illegible]	[illegible]

LAT 36°N

LHA ♈	Hc	Zn	Hc	Zn	Hc	Zn	Hc	Zn	Hc	Zn	Hc	Zn	Hc	Zn
	•DENEB		ALTAIR		Nunki		•ANTARES		ARCTURUS		•Alkaid		Kochab	
270	58 28	060	53 03	131	26 23	166	23 58	202	38 11	267	43 05	307	45 38	343
271	59 10	060	53 40	132	26 34	167	23 39	203	37 22	268	42 27	307	45 24	343
272	59 52	060	54 16	133	26 44	168	23 20	204	36 34	268	41 48	308	45 10	343
273	60 34	060	54 51	135	26 53	169	23 00	205	35 45	269	41 10	308	44 55	343
274	61 16	060	55 25	136	27 02	170	22 39	206	34 57	269	40 31	308	44 41	343
275	61 58	060	55 58	137	27 10	171	22 17	207	34 08	270	39 53	308	44 26	342
276	62 40	060	56 30	139	27 16	172	21 55	208	33 20	271	39 15	308	44 12	342
277	63 22	059	57 02	140	27 22	173	21 32	209	32 31	271	38 37	308	43 57	342
278	64 04	059	57 32	142	27 28	174	21 08	210	31 43	272	37 59	309	43 42	342
279	64 45	059	58 01	144	27 32	175	20 44	210	30 54	272	37 21	309	43 27	342
280	65 27	059	58 30	145	27 36	176	20 19	211	30 06	273	36 43	309	43 12	342
281	66 08	058	58 57	147	27 38	177	19 54	212	29 17	273	36 05	309	42 56	342
282	66 49	058	59 22	149	27 40	178	19 28	213	28 29	274	35 28	309	42 41	341
283	67 30	057	59 47	151	27 41	179	19 01	214	27 40	275	34 50	310	42 25	341
284	68 11	057	60 10	152	27 41	180	18 34	215	26 52	275	34 13	310	42 10	341
	Alpheratz		•Enif		ALTAIR		•Rasalhague		ARCTURUS		Alkaid		•Kochab	
285	26 27	072	44 40	115	60 32	154	59 39	225	26 04	276	33 36	310	41 54	341
286	27 13	073	45 24	116	60 52	156	59 04	226	25 15	276	32 59	310	41 38	341
287	27 59	073	46 07	117	61 11	158	58 28	228	24 27	277	32 22	311	41 23	341
288	28 46	073	46 51	118	61 29	160	57 52	229	23 39	277	31 46	311	41 07	341
289	29 32	074	47 33	119	61 45	162	57 15	231	22 51	278	31 08	311	40 51	341
290	30 19	074	48 16	120	61 59	164	56 37	232	22 03	278	30 32	311	40 35	341
291	31 06	075	48 57	121	62 11	166	55 58	233	21 15	279	29 55	312	40 19	341
292	31 53	075	49 39	122	62 22	168	55 19	235	20 27	279	29 19	312	40 03	341
293	32 40	076	50 20	123	62 31	170	54 39	236	19 39	280	28 43	312	39 46	341
294	33 27	076	51 00	124	62 38	172	53 58	237	18 51	281	28 07	313	39 30	340
295	34 14	077	51 40	125	62 44	175	53 17	238	18 04	281	27 32	313	39 14	340
296	35 01	077	52 19	127	62 48	177	52 35	240	17 16	282	26 56	313	38 58	340
297	35 49	078	52 58	128	62 49	179	51 53	241	16 28	282	26 21	313	38 41	340
298	36 36	078	53 36	129	62 49	181	51 11	242	15 41	283	25 46	314	38 25	340
299	37 24	079	54 13	130	62 48	183	50 28	243	14 54	283	25 11	314	38 09	340
	•Mirfak		Alpheratz		•FOMALHAUT		ALTAIR		Rasalhague		•Alphecca		Kochab	
300	15 17	039	38 11	079	12 16	142	62 44	185	49 45	244	33 34	281	37 53	340
301	15 47	039	38 59	079	12 46	142	62 39	188	49 01	245	32 46	281	37 36	340
302	16 18	040	39 47	080	13 15	143	62 31	190	48 17	246	31 59	282	37 20	340
303	16 49	040	40 35	080	13 44	144	62 22	192	47 32	247	31 11	282	37 04	340
304	17 20	040	41 23	081	14 13	145	62 12	194	46 48	248	30 24	283	36 47	340
305	17 52	041	42 10	081	14 40	145	61 59	196	46 03	249	29 36	283	36 31	340
306	18 24	041	42 58	082	15 08	146	61 45	198	45 17	249	28 49	284	36 15	340
307	18 56	041	43 47	082	15 34	147	61 29	200	44 32	250	28 02	284	35 58	340
308	19 28	042	44 35	083	16 01	148	61 12	202	43 46	251	27 15	285	35 42	340
309	20 00	042	45 23	083	16 26	149	60 53	204	43 00	252	26 28	285	35 26	341
310	20 33	043	46 11	084	16 51	149	60 33	206	42 14	253	25 41	286	35 10	341
311	21 06	043	46 59	084	17 16	150	60 11	208	41 27	254	24 55	286	34 54	341
312	21 39	043	47 48	085	17 40	151	59 48	209	40 40	254	24 08	287	34 38	341
313	22 13	044	48 36	085	18 03	152	59 23	211	39 54	255	23 22	287	34 22	341
314	[illegible]	[illegible]	[illegible]	[illegible]	[illegible]	[illegible]	[illegible]	[illegible]	[illegible]	[illegible]	[illegible]	[illegible]	[illegible]	[illegible]

225	27 41 052	19 18 093	48 12 114	24 12 158	37 48 210	42 19 256	45 30 325
226	28 20 053	20 06 094	48 56 115	24 30 159	37 23 211	41 32 257	45 01 325
227	28 59 053	20 55 094	49 39 116	24 47 160	36 57 212	40 44 258	44 33 324
228	29 37 053	21 43 095	50 23 117	25 03 161	36 31 214	39 57 258	44 05 324
229	30 16 054	22 31 096	51 06 118	25 18 162	36 04 215	39 09 259	43 37 324
230	30 56 054	23 20 096	51 48 119	25 33 163	35 36 216	38 21 260	43 08 324
231	31 35 054	24 08 097	52 31 120	25 46 164	35 07 217	37 34 261	42 40 324
232	32 15 055	24 56 097	53 12 121	25 59 165	34 38 218	36 46 261	42 12 324
233	32 54 055	25 44 098	53 53 123	26 12 166	34 08 219	35 58 262	41 44 324
234	33 34 055	26 32 099	54 34 124	26 23 167	33 37 220	35 10 263	41 15 324
235	34 14 055	27 20 099	55 14 125	26 34 168	33 05 221	34 21 263	40 47 324
236	34 54 056	28 08 100	55 53 126	26 43 169	32 33 222	33 33 264	40 19 324
237	35 34 056	28 56 101	56 32 128	26 52 170	32 01 223	32 45 265	39 50 324
238	36 14 056	29 43 101	57 10 129	27 01 171	31 28 224	31 56 265	39 22 324
239	36 55 057	30 31 102	57 47 130	27 08 172	30 54 225	31 08 266	38 54 324
	♦DENEB	ALTAIR	Nunki	♦ANTARES	ARCTURUS	♦Alkaid	Kochab
240	37 35 057	31 18 103	15 21 140	27 14 173	61 30 241	62 27 309	50 52 353
241	38 16 057	32 06 103	15 52 141	27 20 174	60 47 242	61 49 309	50 46 352
242	38 57 057	32 53 104	16 22 142	27 25 175	60 04 244	61 11 309	50 39 352
243	39 38 057	33 40 105	16 52 142	27 29 176	59 20 245	60 33 308	50 32 351
244	40 19 058	34 27 106	17 21 143	27 32 177	58 36 246	59 55 308	50 25 351
245	41 00 058	35 13 106	17 50 144	27 34 178	57 52 247	59 17 308	50 17 351
246	41 41 058	36 00 107	18 18 145	27 35 179	57 07 248	58 39 308	50 09 350
247	42 22 058	36 46 108	18 46 146	27 36 180	56 22 249	58 00 307	50 01 350
248	43 04 059	37 32 109	19 13 146	27 36 181	55 36 250	57 22 307	49 52 350
249	43 45 059	38 18 109	19 40 147	27 35 182	54 50 251	56 43 307	49 43 349
250	44 27 059	39 04 110	20 06 148	27 32 183	54 04 252	56 04 307	49 34 349
251	45 08 059	39 49 111	20 31 149	27 30 184	53 18 253	55 25 307	49 24 349
252	45 50 059	40 34 112	20 56 150	27 26 185	52 31 254	54 46 307	49 15 348
253	46 32 059	41 19 113	21 20 151	27 21 186	51 45 255	54 07 307	49 05 348
254	47 13 060	42 04 114	21 43 152	27 16 187	50 58 256	53 28 307	48 54 348
	♦DENEB	ALTAIR	Nunki	♦ANTARES	ARCTURUS	♦Alkaid	Kochab
255	47 55 060	42 48 114	22 06 152	27 10 188	50 11 256	52 49 307	48 44 347
256	48 37 060	43 32 115	22 28 153	27 02 189	49 23 257	52 10 306	48 33 347
257	49 19 060	44 16 116	22 50 154	26 54 190	48 36 258	51 31 306	48 22 347
258	50 01 060	44 59 117	23 10 155	26 46 191	47 48 259	50 52 306	48 10 346
259	50 43 060	45 42 118	23 30 156	26 36 192	47 01 260	50 13 306	47 59 346
260	51 25 060	46 25 119	23 50 157	26 26 193	46 13 260	49 34 306	47 47 346
261	52 08 060	47 07 120	24 08 158	26 14 194	45 25 261	48 55 306	47 35 346
262	52 50 060	47 49 121	24 26 159	26 02 195	44 37 262	48 16 307	47 23 345
263	53 32 060	48 30 122	24 44 160	25 50 196	43 49 262	47 37 307	47 10 345
264	54 14 060	49 11 123	25 00 161	25 36 197	43 01 263	46 58 307	46 58 345
265	54 56 061	49 51 125	25 16 162	25 22 198	42 13 264	46 19 307	46 45 345
266	55 39 061	50 31 126	25 31 163	25 06 199	41 24 264	45 40 307	46 32 344
267	56 21 061	51 10 127	25 45 163	24 51 200	40 36 265	45 02 307	46 19 344
268	57 03 061	51 48 128	25 58 164	24 34 201	39 48 266	44 23 307	46 05 344
269	57 45 060	52 26 129	26 11 165	24 16 201	38 59 266	43 44 307	45 51 344

315	23 20 044	23 56 078	14 32 126	18 48 153	58 30 215	61 32 286	33 50 341
316	23 54 045	24 43 078	15 11 126	19 09 154	58 02 216	60 45 287	33 34 341
317	24 29 045	25 31 079	15 50 127	19 30 155	57 33 218	59 59 287	33 18 341
318	25 03 045	26 19 079	16 29 128	19 50 156	57 03 219	59 12 287	33 02 341
319	25 38 046	27 06 080	17 07 129	20 09 157	56 31 221	58 26 287	32 47 341
320	26 12 046	27 54 080	17 45 129	20 28 158	55 59 222	57 40 287	32 31 341
321	26 47 046	28 42 081	18 22 130	20 46 159	55 26 224	56 53 288	32 16 341
322	27 23 047	29 30 081	18 59 131	21 04 159	54 52 225	56 07 288	32 00 342
323	27 58 047	30 18 082	19 35 132	21 20 160	54 17 227	55 21 288	31 45 342
324	28 33 047	31 06 082	20 11 133	21 36 161	53 41 228	54 35 288	31 30 342
325	29 09 047	31 54 083	20 47 133	21 52 162	53 04 229	53 49 289	31 15 342
326	29 45 048	32 42 083	21 22 134	22 06 163	52 27 231	53 03 289	31 00 342
327	30 21 048	33 31 084	21 56 135	22 20 164	51 49 232	52 17 289	30 45 342
328	30 57 048	34 19 084	22 31 136	22 33 165	51 11 233	51 31 289	30 30 342
329	31 34 049	35 07 085	23 04 137	22 46 166	50 32 234	50 45 290	30 15 342
	♦CAPELLA	Hamal	Diphda	♦FOMALHAUT	ALTAIR	♦VEGA	Kochab
330	13 56 043	35 56 085	23 37 137	22 57 167	49 52 235	49 59 290	30 01 343
331	14 29 043	36 44 086	24 10 138	23 08 168	49 12 237	49 14 290	29 46 343
332	15 02 044	37 32 087	24 42 139	23 18 168	48 31 238	48 28 290	29 32 343
333	15 35 044	38 21 087	25 13 140	23 28 169	47 50 239	47 43 291	29 17 343
334	16 09 044	39 09 088	25 44 141	23 36 170	47 08 240	46 57 291	29 03 343
335	16 43 045	39 58 088	26 15 142	23 44 171	46 26 241	46 12 291	28 49 343
336	17 18 045	40 46 089	26 44 143	23 51 172	45 43 242	45 27 291	28 35 343
337	17 52 046	41 35 089	27 14 144	23 57 173	45 00 243	44 42 292	28 22 344
338	18 27 046	42 23 090	27 42 145	24 02 174	44 17 244	43 57 292	28 08 344
339	19 02 046	43 12 091	28 10 145	24 07 175	43 33 245	43 12 292	27 55 344
340	19 37 047	44 01 091	28 37 146	24 11 176	42 49 246	42 27 293	27 41 344
341	20 13 047	44 49 092	29 04 147	24 14 177	42 05 246	41 42 293	27 28 344
342	20 48 048	45 38 092	29 29 148	24 16 178	41 20 247	40 57 293	27 15 345
343	21 24 048	46 26 093	29 54 149	24 17 179	40 35 248	40 13 294	27 02 345
344	22 00 048	47 15 094	30 19 150	24 18 180	39 50 249	39 28 294	26 49 345
	♦CAPELLA	ALDEBARAN	Diphda	♦FOMALHAUT	ALTAIR	♦VEGA	Kochab
345	22 37 049	14 32 080	30 43 151	24 18 181	39 05 250	38 44 294	26 37 345
346	23 13 049	15 20 081	31 05 152	24 17 182	38 19 251	38 00 294	26 24 345
347	23 50 049	16 08 081	31 28 153	24 15 183	37 33 251	37 16 295	26 12 345
348	24 27 050	16 56 082	31 49 154	24 12 184	36 47 252	36 32 295	26 00 346
349	25 04 050	17 44 082	32 10 155	24 09 185	36 01 253	35 48 295	25 48 346
350	25 42 050	18 32 083	32 29 157	24 04 186	35 14 254	35 04 296	25 37 346
351	26 19 051	19 20 083	32 48 158	23 59 186	34 28 254	34 20 296	25 25 346
352	26 57 051	20 09 084	33 06 159	23 53 187	33 41 255	33 37 297	25 14 347
353	27 35 051	20 57 084	33 24 160	23 47 188	32 54 256	32 54 297	25 02 347
354	28 13 052	21 45 085	33 40 161	23 39 189	32 07 257	32 10 297	24 51 347
355	28 51 052	22 34 086	33 55 162	23 31 190	31 19 257	31 27 298	24 41 347
356	29 29 052	23 22 086	34 10 163	23 22 191	30 32 258	30 44 298	24 30 347
357	30 08 053	24 10 087	34 24 164	23 12 192	29 45 259	30 01 298	24 19 348
358	30 46 053	24 59 087	34 37 165	23 02 193	28 57 259	29 19 299	24 09 348
359	31 25 053	25 47 088	34 48 166	22 51 194	28 09 260	28 36 299	23 59 348

Index

Abbreviations, 107-10
Air Almanac: for planning, 41; source, 93
Almanac. See Air Almanac, Nautical Almanac, Reed's Almanac
Almanac data, wired into computers, 29, 97
Altitude, celestial: defined, 107; negative, 87-89
Altitude, computed (Hc), defined, 108
Altitude, observed (Ho), defined, 108
Altitude, sextant (hs): defined, 108; negative 87-89
American Practical Navigator, source, 94
AP618. *See* H.O. 218
AP3270. *See* H.O. 249
Approximations: for length of day, 27, 60-61; for position at sunrise, sunset, and twilight, 59
Aries (♈), defined, 107
Assumed altitude method. *See* Davies tables
Assumed position (AP), defined, 107
Averaging: to reduce errors, 36-37; of speeds, 48, 49
Azimuth (Zn): defined, 110; formula for, 101; in H.O. 249, 12
Azimuth angle (Z), defined, 110

Barometer, aneroid, calibration, 22
Bayless tables, *Compact Sight Reduction Table (Modified H.O. 211, Ageton's Table):* as backup, 12; comparison, 96
Bearings to check compass, 16
Blunders, avoiding, 34
Board of Longitude and £20,000 prize, 23, 24
Bowditch, source, 94
Broadcasts, radio: for landfall, 77-78; weather, 17-21, 78, 93

Calculator: for backup sight reduction, 12; for interpolation, 59-60; for planning, 42
Calculator, engineering (slide-rule): for backup sight reduction, 29, 98; cost, 29; navigation uses, 29
Calculator, programmable: for celestial identification, 32-33; celestial-navigation program, 30-33; compared to H. O. 249, 29-30, 98; entering and reading degrees, minutes, and tenths, 30-32; for great-circle calculations, 29-30, 50; for interpolation (using linear regression), 102; for sight reduction, 30-33
Calculator, specialized navigation, 29, 97-98
Celestial bodies available for sighting, 72
Celestial navigation, need for, 5-6
Celestial sights: in cloudy weather, 76; evening, 74; morning, 73, 75; at night, body on horizon, 5, 85-89; repeating, 75-76
Changes in plan, 70
Checklists: navigation equipment, 99-100; navigator's preparations, 100; survival equipment, 100

Chronometer: invention of, 23-24; superseded by quartz watch, 24
Cloudy weather, celestial sights during, 76
Compass: hand-bearing (to check steering compass), 16; metal near, 14; swinging, 15-16
Course to steer, predicted, 56-57
Current(s): allowing for (triangle of forces), 56-57; predicting, 48, shown on pilot charts, 42

Daily navigation guides, examples, 73, 74, 75
Davies tables: comparison, 29, 79-80, 97; knowlege of, 5; in planning, 41, 80-81; unique feature of, 79; use of, 79-81
Days: keeping straight, 69; length of, 25-26
d correction for minutes and seconds of time in *Nautical Almanac*, 107
d correction for minutes of declination, 107
Dead reckoning (DR): to check celestial fixes, 38; definition, 107; importance at sea, 38-39
Declination, defined, 107
Defense Mapping Agency, for publications, 92-94
Departure time: optimum, 42, 47-48; uncertain, 61-62
Differencing to reveal errors, 35-36
Dip, defined, 107
Direction finder. *See* Radio direction finder
Double sideband, WWV broadcasts in, 19
DR. *See* Dead reckoning

Electronic equipment: checking before departure, 15; failures, 6-7; protecting, 6, 15; substitutes for, 6-7
Emergency equipment, 100
Equation of time: defined, 108; for noonsight, 57, 83; rules for applying, 102
Equipment: alternatives, 6; emergency, 100; preparations, 14-16; protecting, 6, 15
Errors: averaging to reduce, 36-37; avoiding, 33-34; checking for, 37-38; differencing to detect, 35-36; in fixes, 37-38; reducing, 34
Estimating: average heading, 39; leeway, 39; speed, 39-40
Evening sights, daily guide for, 74
Eye patch for celestial sights, 15

Fast plan, worksheets for, 69-70
Fix: four or more LOPs preferable, 37; sun-moon, 85
Form for sight reduction, 90-91
Formulas: azimuth angle, 101; equation of time, 102; interpolation (by linear regression), 102; noonsight, 101; sight reduction (sine-cosine), 101; speed (average), 48-49; speed (estimated), 40; storm avoidance, 104-6; true wind, 102-4
Frequency indicator for tuning radio, 18

Gales, predicted, on pilot charts, 42-47. *See also* Storm(s)
Geographical position (GP): defined, 108; for high-altitude sights, 84
GHA. *See* Greenwich hour angle
GHA ♈. *See* Greenwich hour angle of Aries
GMT. *See* Greenwich mean time
Gnomonic chart for great-circle course, 49
Government Printing Office, U.S., for publications, 92-94
GP. *See* Geographical position

Great-circle course: computing by calculator, 29-30, 49-50; computing by H.O. 211, H.O. 229, H.O. 249 Vols. II and III, 49-50; vs. rhumb line, 49-50
Greenwich hour angle (GHA), defined, 108
Greenwich hour angle of Aries (GHA ♈), for rising and setting of planets, 64-68
Greenwich mean time (GMT): defined, 108; interpolation for, 68; vs. local zone time (LZT) for navigation, 25

Harmonic mean: formula for averaging speeds, 48-49; vs. simple average, 52
Harrison, John, inventor of chronometer, 24
Hc. *See* Altitude, computed
HD486. *See* H.O. 214
Heading: best, for storm avoidance, 104-5; estimating average, 39
Height of eye (HE): defined, 108; effect of waves on, 14-15; measuring, 14. *See also* Dip
High-altitude sights: of sun, 84; Table of Offsets (H.O. 229) for correcting, 84
Ho. *See* Altitude, observed
H.O. 208, *Navigation Tables:* as backup method, 12; comparison, 96-97
H.O. 211, *Dead Reckoning Altitude and Azimuth Table:* Bayless modification, 12; comparison, 96; for great-circle computations, 49
H.O. 214, *Tables of Computed Altitude and Azimuth,* comparison, 96
H.O. 218, *Astronomical Navigation Tables,* comparison, 97
H.O. 229, *Sight Reduction Tables for Marine Navigation:* vs. calculator, 29; comparison, 96; for great-circle computations, 49-50; knowledge of , 4-5; for negative altitudes, 88; source, 93; use of, 12
H.O. 249, *Sight Reduction Tables for Air Navigation:* accuracy, 12; vs. calculator, 29-30; comparison, 96; contents, 11-12; for great-circle computations, 49-50; knowledge of, 4-5; for negative altitudes, 88; source, 93; for star identification (Vol. I), 13; for twilight planning (Vol. I), 72; use of, 11-12
H.O. 2101. *See* Rude's Star Finder and Identifier
Horizon: sights when not visible, 85-89; sights with body on, 85-89
Horizontal parallax (HP), defined, 108
hs. *See* Altitude, sextant
Hurricanes: avoiding, 18; danger in summer, 43-45; predicted, shown on pilot charts, 42-47

Icebergs, range in July, 49
Index correction (IC), sextant: checking, 14; defined, 108
Intercept, defined, 108
Interpolation: with a calculator, 102; by eye, 60; for GMT, 68; for moonrise and moonset, 63; in *Nautical Almanac,* for times of sunrise, sunset, and twilight, 59-60; for worksheet data, 61; on worksheet for rising and setting of planets, 68

Jupiter. *See* Planets

LAN. *See* Local apparent noon

Landfall: Bermuda example, 78; planning for, 77-78; using single LOP for, 77-78
Lands End to New York, cruise planning, 42-45
LAT. *See* Local apparent time
Leeway, estimating, 39
Length of the day, not always 24 hours, 25-26
LHA ♈. *See* Local hour angle of Aries
Lifeboat: checklist for equipment, 100; equipment, 16; sight reduction for, 12
Light Lists, U. S. Coast Guard, source, 94
Lights for landfall, 77-78
Linear regression for interpolation, 102
Line of position (LOP): four or more preferred, 37; single, for landfall, 77
List of Lights, source, 94
LMT. *See* Local mean time
Local apparent noon (LAN): defined, 109; expected position at, 26-27; projected altitude at, 84; time of, 26, 57, 61, 83
Local apparent time (LAT), defined, 109
Local boat time, for daily activities, 25
Local hour angle (LHA), defined, 109
Local hour angle of Aries (LHA ♈): on daily navigation guides, 72-75; in H.O. 249, 12; to set star finder, 72
Local mean time (LMT), defined, 109
Local zone time (LZT): defined, 109; vs. GMT for navigation, 25
Longitude: correction, for moonrise and moonset, 62; need to determine in 18th century, 23-24; time equivalent of, 57, 61-63
LOP. *See* Line of position
Loran, limitations, 5-6
Lower limb (LL), corrections, defined, 109
LZT. *See* Local zone time

Mars. *See* Planets
Moon: minimum distance from sun for observation, 65, 76; planning sights, 7; predicted rising and setting times, 73, 75; sights at twilight, 76
Moonrise and moonset, times of, 62-63
Moon-sun fixes, 85
Morning sights, daily guide for, 73, 75
Morse code: learning, 21; recording, 21; weather broadcasts in, 18, 20-21

Nautical Almanac: accuracy of sunrise, sunset, and twilight times, 36, 61; equation of time in, 57; interpolation in, 59-60; for moonrise and moonset, 62-63; for planning, 41; rounding figures from, 12; source, 92-93; for times of LAN, sunrise, sunset, and twilight, 57-60
Nautical mile (N.M.), defined, 109
Navigator: advantage of being, 10; job of, 8-10; responsibility of, 10; standing watches, 8; work schedule of, 8-9
NAVSAT, limitations of, 5-6
Negative numbers, dealing with, 88-89
Night, sights during, 5, 85-89. *See also* Sights at night
Noonsight of the sun: averaging for, 83; formulas, 101; high-altitude sights for, 84; in history, 83; planning, 7, 83; projected time and altitude for, 73, 75, 83-84; value of, 83

Norfolk-Bermuda cruise planning, 45-53
NP401. *See* H.O. 229

Ocean Passages for the World: for planning, 41; route from Lands End to New York, 44; source, 92
Offsets, table of (H.O. 229), for high-altitude sights, 84
Omega, limitations, 5-6
Order of sighting bodies at twilight, 72

Parallax (P), defined, 109
Pilot charts: excerpts from, 46-47; for planning, 41, 42-48; source, 92; summarizing data from, 45
Plan: changes in, 70; slow vs. fast, 53
Planets: identification of, 13; minimum distance from sun for observation, 64-65; planning sights of, 7; rising and setting of, 64-69; star finder and, 64
Planning: deviations from, 7-8; elements of, 7; for first day of cruise, 57, 59, 60; purpose of, 4, 5; revisions in, 7-8; simplifications, 61; time required for, 5; value of, 3-4
Position, expected, for planning, 26-27, 57, 59-61
Precession and nutation, correction for, 80
Predicted altitudes and azimuths at twilight, 72-75
Preparations for cruise, 14
Publications: non-Government, 95; U.S. Government, 92-94
Pub. No. 229. *See* H.O. 229
Pub. No. 249. *See* H.O. 249

Radio Aids to Marine Navigation, source, 94
Radio broadcasts: for landfall, 77-78; of time signals, 13-14, 24; of weather reports, 17-21, 78, 93
Radio direction finder (RDF): corrections for, 15-16; metal loops as cause of errors in, 15
Radio Navigational Aids, source, 93
Radio transmitters for landfall, 77-78
RDF. *See* Radio direction finder
Reed's Almanac: alternative for *Nautical Almanac,* 12; for planning, 41
Refraction (R), defined, 109
Regression, linear, for interpolation, 102
Relative movement in storm avoidance, 106
Repeating celestial sights for accuracy, 75-76
Rhumb line vs. great circle, 49, 50
Right ascension (RA): defined, 109; for star finder, 64, 65, 66, 67
Rounding off numbers: recommended method, 35; in sight reduction, 12
Rude's Star Finder and Identifier, H.O. 2102: for identification, 76; knowledge of, 5; for planning, 41, 72; preparing, 64; for rising and setting of planets, 64-69

Sailing ship routes, in *Ocean Passages for the World,* 44
Saturn. *See* Planets
Selected Worldwide Marine Weather Broadcasts: use in planning, 18, 21; source, 93
Semidiameter (SD), defined, 110
Sextant: adjusting, 14; backup, 16; checking, 14; cleaning, 14; emergency, 100
Sextant altitude. *See* Altitude, sextant
SHA. *See* Sidereal hour angle
Shipping lanes shown on pilot charts, 18, 42

charts, 18, 42
Shortcuts: for length of day, 27, 60-61; for position at sunrise, sunset, and twilight, 59
Side error, sextant, checking and adjusting, 14
Sidereal hour angle (SHA), defined, 110
Sighting order, bodies at twilight, 72
Sight reduction: backup methods, 12; form, 90-91; formulas for Hc and Z, 101; methods compared, 96-98; for negative altitudes, 87-89; practice, 13; recommended methods, 11-12
Sight Reduction Tables for Air Navigation, Pub. No. 249. *See* H.O. 249
Sight Reduction Tables for Marine Navigation, Pub. No. 229. *See* H.O. 229
Sights at night, body on horizon, 5, 74, 77, 85-89
Silica gel to protect equipment, 15
Simplifications: for length of day, 27, 60-61; for position at sunrise, sunset, and twilight, 59
Sine-cosine formula: comparison, 98; equation, 101
Single LOP for landfall, 77
Single sideband (SSB) broadcasts, tuning in, 18
Speed: allowing for current, 56-57; estimating, 6-7, 39-40, 51-52; formula for averaging, 48-49; formula for estimating, 40
Speed made good (SMG): defined, 110; SMG factor (cosine), 52; when tacking, 51-52, 56-57
Star finder. *See* Rude's Star Finder and Identifier
Star identification: methods, 13; practice, 13
Static, radio: for predicting thunderstorms, 21-22; related to frequency, 22
Steamship routes, shown on pilot charts, 18, 42
Steering: automatic, 39; star for guide, 39
Storm(s): avoidance formula, 105; avoiding, 18-20; while awaiting departure, 18, 20
Sun, positions marked on star finder, 69. *See also* Noonsight
Sun-moon fixes, 38, 85
Sunrise. *See Nautical Almanac,* Twilight
Sunset. *See Nautical Almanac,* Twilight
Survival equipment, checklist, 100

Tacking, effect on speed made good (SMG), 51-52
Temperature, air and sea, predicted on pilot charts, 42
Thunderstorm predictions, radio static as indication, 21-22
Timekeeping at sea, LZT vs. GMT, 25
Timepiece: error, 13-14; quartz digital, 24; rating, 13-14; recommended for navigation, 24; setting to GMT, 24; setting to local boat time, 24; three for reliability, 24
Time signal, 13-14, 24
Triangle of forces: for effect of current, 56-57; for storm avoidance, 105-6; for true wind, 102-4
Trigonometric tables, as backup for sight reduction, 12, 98, 101
True course (TC) to steer, predicted, allowing for current, 56-57

Twilight: accuracy of *Nautical Almanac* data for, 36, 61; interpolation for time of, 59-60; planning for, 71-77; position at, 26; predicted time of, 26, 57-61, 71-77; time to begin sights at, 73-75

Upper limb (UL) corrections, defined, 110

v correction, defined, 110
Vector diagrams. *See* Triangle of forces
Venus. *See* Planets

Watch, navigation. *See* Timepiece
Watch routine and navigator, 8-9
Waves: effect on dip, 14-15; predicted, shown on pilot charts, 42-47
Weather: altering course for, 18; broadcasts, 17-21, 78, 93; knowledge of, 6; planning for, 17-22; thunderstorms, 21-22. *See also* Gales, Hurricanes, Storm(s)
Wind(s): finding true from apparent, 102-4; prevailing, shown on pilot charts, 42-47
Worksheets: for fast plan, 69-70; for slow plan, 58, 63, 65, 66-67; steps in preparation, 55-69
WWV: time signal, 13-14, 24; weather broadcasts, 18-20

Z. *See* Azimuth angle
Zn. *See* Azimuth